自然与建筑
Nature and Architecture
Wang Yun
王昀

中国电力出版社
CHINA ELECTRIC POWER PRESS

图书在版编目（CIP）数据

跨界设计：自然与建筑 / 王昀著. — 北京：中国电力出版社，2017.1
ISBN 978-7-5123-9902-0

Ⅰ.①跨… Ⅱ.①王… Ⅲ.①建筑设计 Ⅳ.①TU2

中国版本图书馆CIP数据核字(2016)第247178号

感谢北京建筑大学"研究生教学研究与质量提升项目（2016年）"的经费支持

内容提要

从自然形成的地形地貌中截取相应的形态，并以此进一步地对形态本身进行空间层面的观察与操作是该书的主旨。通过对所选取的地形进行空间图式的抽取与建模，进而整理出由地形所产生的拥有建筑含义的空间形态，其过程本身不仅仅是对于空间的观察，其所做的一切，事实上也是一种对于所谓地域性问题的描写与诠释。从俯瞰的视点对"地域性"表达进行新的思考和观测是本书的主要特质。本书适合从事建筑设计、研究及评论者阅读。

中国电力出版社出版发行
北京市东城区北京站西街19号 100005
http://www.cepp.sgcc.com.cn
责任编辑：王 倩
封面设计：方体空间工作室（Atelier Fronti）
版式设计：王风雅 姚博健
责任印制：蔺义舟
责任校对：王开云
英文翻译：陈伟航
北京盛通印刷股份有限公司印制·各地新华书店经售
2017年1月第1版·第1次印刷
710mm×980mm 1/16·11印张·2插页·198千字
印数：1—1800册
定价：58.00元

Abstract

The theme of this book is to extract relevant patterns from landforms formed naturally, then observe and process such patterns from a spatial perspective. Through spatial schematic extraction and modeling based on the selected landforms, we further sort out spatial patterns with architectural meanings originated from landforms. This process itself is more than a form of observation on space. In fact, all it involves is a kind of description and interpretation on the issue of so-called regionalism. The major feature of this book is to conduct new contemplation and observation on the expression of "regionalism" from a top-to-bottom overseeing perspective. It is suitable for architectural designers, researchers and commentators to read this book.

敬告读者
本书封底贴有防伪标签，刮开涂层可查询真伪
本书如有印装质量问题，我社发行部负责退换
版权专有 翻印必究

序 Preface

　　将大自然的地貌形态直接转化为具有建筑含义的空间形态，是这本书重点要讨论的问题。这种将自然与建筑相结合的想法，源于20世纪90年代末的一次从北京至成都的飞机途中。当我从由北京出发的班机上俯瞰到太原一带如刀削的地貌，俯瞰到进入西安一带的高原台地，俯瞰到四川境内烟雾缭绕的山峦，过程中所呈现的山形地貌的变化以及彼此间的对比，都在述说着地域本身所拥有的视觉特征。地形的画面感，河流的曲线，让我看到了其中所呈现的空间，看到了其中所呈现的建筑，自然也看到了展现于其表面的地形地貌所呈现出的体块特征。

　　随着卫星地图的发展，"鸟瞰自然"本身变得如此容易，获得大地图像的方式也变得如此轻而易举，而这种"鸟瞰自然"图像技术的发展最终也促使着观念的变化。

　　于此所呈献的这本《建筑与自然》，是希望能够将我在这一系列"鸟瞰自然"过程中所看到的大地这一自然形态本身所呈现出的空间与建筑转而呈献给读者，希冀透过这一切，使读者能够由此而看到和发现更多自然本身正在呈现着的新的空间与建筑的世界。

The book focuses on the issue of direct transformation from natural landforms into spatial patterns with architectural meanings. This idea of integrating nature with architecture originated from a flight from Beijing to Chengdu in the late 1990s. Starting from Beijing, I watched downwards from the plane the difference and contrast among various landforms along the journey, such as steep cliffs around Taiyuan city, plateaus and terraces around Xi'an city and mist-shrouded mountains in Sichuan province, all expressing visual features possessed by territories themselves. The picturesque landforms and curvy rivers, in addition to presenting block features on the surface, reveal space patterns and architecture forms contained within.

With the development of satellite maps, it becomes easier and easier to "get a bird's-eye view of nature" and obtain images of land. Meanwhile, the development of such graphic technology may finally promote change of concepts.

This book, *Nature and Architecture*, is expected to show readers what I have discovered during serial observations of "aerial view of nature": space and architecture revealed by natural patterns of land. I hope readers can, through all these, see and find more new spaces and architectures worlds that nature itself is revealing.

王昀
Wang Yun
2016年05月

目录

序
导读

1 非"透视学"视点下的"写生"   1

2 从地形到空间的 19 个实例   7
    实例 2-1   从"北纬 23.45'36.63",东经 117.22'11.19""周边地形所选择出的建筑   8
    实例 2-2   从"北纬 23.54'55.90",东经 57.06'02.58""周边地形所选择出的建筑   14
    实例 2-3   从"北纬 31.52'20.67",东经 36.58'39.98""周边地形所选择出的建筑   18
    实例 2-4   从"北纬 31.53'02.68",东经 36.52'23.09""周边地形所选择出的建筑   24
    实例 2-5   从"北纬 32.11'20.17",东经 36.08'46.05""周边地形所选择出的建筑   30
    实例 2-6   从"北纬 23.45'01.54",东经 117.14'03.03""周边地形所选择出的建筑   36
    实例 2-7   从"北纬 32.13'03.67",东经 36.08'35.01""周边地形所选择出的建筑   44
    实例 2-8   从"北纬 32.13'10.41",东经 36.14'03.79""周边地形所选择出的建筑   52
    实例 2-9   从"北纬 23.54'27.80",东经 117.28'36.57""周边地形所选择出的建筑   60
    实例 2-10   从"北纬 32.06'07.90",东经 37.22'11.49""周边地形所选择出的建筑   66
    实例 2-11   从"北纬 14.56'29.75",东经 43.23'25.95""周边地形所选择出的建筑   74
    实例 2-12   从"北纬 14.52'56.87",东经 43.17'50.12""周边地形所选择出的建筑   80
    实例 2-13   从"北纬 32.16'08.40",东经 36.05'48.26""周边地形所选择出的建筑   84

实例 2-14	从"北纬 53.50'51.13"，东经 73.24'12.32""周边地形所选择出的建筑	90
实例 2-15	从"北纬 34.17'24.41"，东经 57.33'20.18""周边地形所选择出的建筑	96
实例 2-16	从"北纬 39.59'50.83"，东经 120.43'55.90""周边地形所选择出的建筑	104
实例 2-17	从"北纬 37.30'33.17"，东经 122.00'32.90""周边地形所选择出的建筑	110
实例 2-18	从"北纬 35.38'12.54"，东经 120.46'43.01""周边地形所选择出的建筑	116
实例 2-19	从"北纬 40.45'16.15"，东经 144.52'45.92""周边地形所选择出的建筑	122

3 从自然中直接截取的建筑　　129

实例 3-1	直接截取"北纬 36.75'84.63"，东经 110.49'35""周边隆起的地形所获得的空间形态	130
实例 3-2	直接截取"北纬 37.46'65.09"，东经 110.99'91.12'""周边隆起的地形所获得的正向空间形态	136
实例 3-3	直接截取"北纬 37.46'65.09"，东经 110.99'91.12'""周边隆起的地形所获得的负向空间形态	142
实例 3-4	直接截取"北纬 24.38'16.94"，东经 121.11'84.74'""周边隆起的地形所获得的空间形态	148
实例 3-5	直接截取"北纬 7.32'52.19"，东经 134.49'15.70'""周边隆起的地形所获得的空间形态	152

Contents

Preface

Introduction

1 "Paint from life" based on "non-perspective science"   1

2 19 samples of transformation from landform to space   7

Sample 2-1	Architecture chosen from landform around 23.45'36.63" N, 117.22'11.19" E	8
Sample 2-2	Architecture chosen from landform around 23.54'55.90" N, 57.06'02.58" E	14
Sample 2-3	Architecture chosen from landform around 31.52'20.67" N, 36.58'39.98" E	18
Sample 2-4	Architecture chosen from landform around 31.53'02.68" N, 36.52'23.09" E	24
Sample 2-5	Architecture chosen from landform around 32.11'20.17" N, 36.08'46.05" E	30
Sample 2-6	Architecture chosen from landform around 23.45'01.54" N, 117.14'03.03" E	36
Sample 2-7	Architecture chosen from landform around 32.13'03.67" N, 36.08'35.01" E	44
Sample 2-8	Architecture chosen from landform around 32.13'10.41" N, 36.14'03.79" E	52
Sample 2-9	Architecture chosen from landform around 23.54'27.80" N, 117.28'36.57" E	60
Sample 2-10	Architecture chosen from landform around 32.06'07.90" N, 37.22'11.49" E	66
Sample 2-11	Architecture chosen from landform around 14.56'29.75" N, 43.23'25.95" E	74
Sample 2-12	Architecture chosen from landform around 14.52'56.87" N, 43.17'50.12" E	80
Sample 2-13	Architecture chosen from landform around 32.16'08.40" N, 36.05'48.26" E	84
Sample 2-14	Architecture chosen from landform around 53.50'51.13" N, 73.24'12.32" E	90
Sample 2-15	Architecture chosen from landform around 34.17'24.41" N, 57.33'20.18" E	96

Sample 2-16	Architecture chosen from landform around 39.59'50.83" N, 120.43'55.90" E	104
Sample 2-17	Architecture chosen from landform around 37.30'33.17" N, 122.00'32.90" E	110
Sample 2-18	Architecture chosen from landform around 35.38'12.54" N, 120.46'43.01" E	116
Sample 2-19	Architecture chosen from landform around 40.45'16.15" N, 144.52'45.92" E	122
3 Architecture directly extracted from nature		129
Sample 3-1	Spatial patterns acquired by directly extracting upheaval area around 36.75'84.63"N, 110.49'35" E	130
Sample 3-2	Spatial patterns of forward direction acquired from upheaval landform around 37.46'65.09" N, 110.99'91.12" E	136
Sample 3-3	Spatial patterns of negative direction acquired from upheaval landform around 37.46'65.09" N, 110.99'91.12" E	142
Sample 3-4	Spatial patterns acquired by directly extracting upheaval area around 24.38'16.94"N, 121.11'84.74" E	148
Sample 3-5	Spatial patterns acquired by directly extracting upheaval area around 7.32'52.19"N, 134.49'15.70" E	152

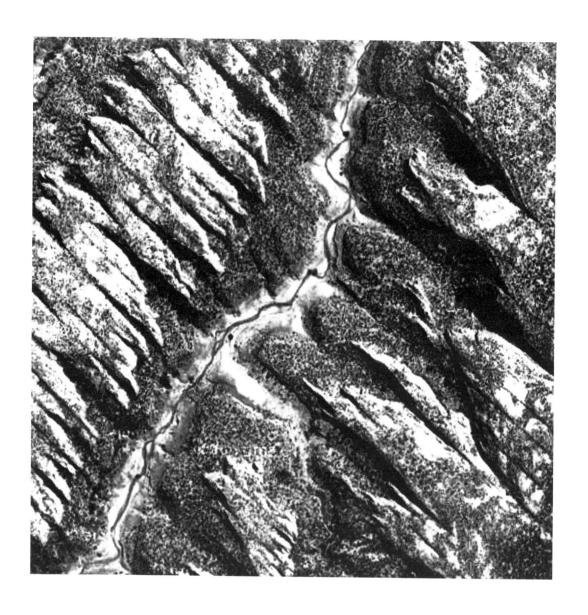

大自然本身所呈现的"肌理"其实已诉诸其平面与空间性的特征
The "texture" revealed by nature actually already resorts to its features of plane and space

导读 Introduction

　　自然的概念具有广泛的含义，动物、植物以及非人为的物质世界均包含在其中。然而在我们的实际生活中，自然本身尽管涉及上述诸多范畴，其范畴本身甚至还包含天气和地质，但由于在更细的分类中，天文、地理、动物、植物等似乎可以再次划分为各自专门的分类，因此本书中所指的自然，实际上是指我们普通人的日常理解，确切地讲，是与我们自己所处的人工世界相对应的自然环境，即一般意义上的由山石、树木与河流所构成的环境。而一旦以这样狭隘的含义与视角去看待自然这个概念，卫星地图上所呈现的城市与乡村之外的区域或许都可以称为"自然"。由于卫星地图上所呈现的"自然"是一种"肌理"的表现，而"肌理"再度呈现出的表象，诉诸印象的又是一种"图案"，而一旦以如此"图案"的视点去审视大自然中所呈现出的表象，大地上所呈现的"肌理"，其实就是一幅幅"抽象绘画"，同时也可以被看作为一个又一个能够诱发空间的要素。

　　自然环境本身会产生相应的"地域特征"，但我们目前所讲的地域性其实只是在"平面"视点上所展开的一种讨论，换言之是人以其活动范围为视野，囿于"透视"视域范围所看到的地域性。然而当人站到空中，从空中俯瞰地形地貌所诉诸的肌理时，或许一种更为广阔的地域整体所表达出的地域性特征便呈现在眼前了。

　　我们在这本书里所强调的地域性，不是那种"不识庐山真面目，只缘身在此山中"的狭隘地域视野所观察到的地域性，我们所说的地域性，是一种"宇宙"视点下所观测到的地域性，是一种辽阔并具有开放性的地域性，而这种广袤与开放的地域性视点，对未来有明确的指向。

　　鉴于这样的思考，本书对自然本身所呈现的，拥有宇宙视点地域性特征与地形地貌表象进行了一系列空间化与建筑化的处理。书中所列举的24个从卫星地图上选择出来的形态，以及从空间的视角对其所进行的空间层面的观察，是一种具体地从世界范围内选择"自然"后并对其进行空间变化与呈现的观测结果。而所有这些源于当地自然原型的新空间的对象物本身，应该说其本身拥有那个地区的地域性。

The concept of nature contains extensive meanings, including animals, plants and material world which is not created by human beings. In reality, nature itself involves not only the various categories mentioned above but also even climate and geology, however, in more detailed classification it seems that astronomy, geology, zoology and botany may further be classified into their own specific categories. Therefore nature indicated by this book actually refers to natural environment we ordinary people understand in everyday life. To be specific, it is natural environment in general sense constituted by mountains, rocks, trees and rivers, in contrast to man-made world we stay in. Once we use this narrow meaning and perspective to view the concept of nature, we may give the name of "nature" to areas appearing on satellite maps other than cities and villages. This "nature" on satellite maps is represented by a kind of "texture", while the phenomenon displayed again by this "texture" is a kind of "pattern" to one's impression. If we use this perspective of "pattern" to examine surface phenomenon displayed by nature, we may find that "textures" on land are actually like "abstract paintings" and can also be viewed as various elements to trigger sense of space.

Natural environment itself can generate corresponding "regional features", but so far regionalism in our discussion actually is merely based on "two-dimensional" perspective. In other words, this regionalism is observed within one's perspective field based on the activity range. However, when a man jumps into the sky and looks downwards at textures presented by landforms, he may see regionalism expressed by broader and vaster territory.

The regionalism we emphasize in this book, instead of that observed by narrow and restricted vision, is the regionalism observed by cosmic vision with broadness and openness, which has a clear and distinctive direction for future.

Based on such thoughts, this book has conducted a series of spatial and architectural processing on regional features and landform phenomenon with cosmic perspective embodied by nature. Listed in this book, the 24 patterns selected from satellite maps and observation with spatial perspective all serve as a result based on spatial change and presentation of "nature" chosen from the world. All objects themselves in new space originated from natural prototypes are expected to possess regionalism of that particular region.

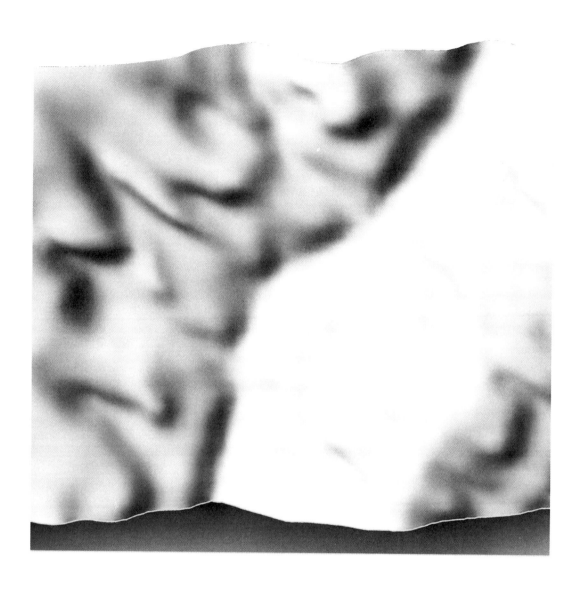

从自然中直接截取的地形所呈现的形态应该拥有那个地区的地域性特征
Patterns presented by landforms directly extracted from nature are expected to possess regionalism of that particular region

自然环境本身诉诸并呈现着无尽的抽象形态
Natural environment itself expresses and presents infinite abstract patterns

1

非"透视学"视点下的"写生"
"Paint from life" based on "non-perspective science"

非"透视学"视点下的"写生"
"Paint from life" based on "non-perspective science"

"写生"是直接以实物或风景为对象的绘画,这种绘画对于学习美术和学习建筑的人来说往往又是一门要学习的必修课程。但是,在对"对象物"进行"写生"时,"对象物"是什么的问题其实是一个很重要的问题,而且与此同等重要的,实际上还有观察世界的视点与观察对象物的角度问题。

以往的"写生"有一个易被忽略的点,那就是写生其实经常被囿于"正常"的视点范围。如人的视线高度通常是根据人的身高在1.7米左右而设定为1.6米左右,而这样的一种日常的视线高度,使得眼睛所看到的是一个由近至远的世界。由于这样的视界所呈现的是近处的物体大,远处的物体小,于是自文艺复兴时期发现"近大远小"原理后,这种"透视"的规则便成为画面整体的构成法则。而这种"透视"的观念,一直影响着人对世界的判断与理解,自然也影响着对写生对象物的判断与理解。大量的绘画遵循并呈现着由近向远看的一层一层的层叠式状态,当你要描述这样的层叠状态时,就需要一层一层走过去。而当"近大远小"的"透视"的形

"Paint from life" depicts objects such as material entities or scenes in reality, which is usually a compulsory course for students of fine arts and architecture. However, it is a very significant issue during depiction to understand what the object is, and it is equally important regarding the viewpoint to observe the world and perspective to observe the object.

Previously "paint from life" tends to ignore one point: the viewpoint is often limited within the so-called "normal" range. The height of sighting line, based on body height of 1.7 meters, is generally set at about 1.6 meters. This everyday height of sighting line makes our eyes see a world from close by to far away, leading to a principle in which nearby objects seem large and distant objects small. With this principle discovered during the Renaissance, the rule of "perspective" becomes the overall constitution of graphics. It affects people's judgment and understanding not only of the world but also of objects to be depicted. Many paintings follow and embody the status of various layers from close by to far away, and you need to walk through all these layers to describe such a stratified status of layers. When such a pattern of perspective is taken for granted and considered as a must, even if you look down from a mountain top, actually you still see a scene

态成为理所当然和必然时，即便你到了山顶，并从山顶上向下看的时候，其实你所看到的也是一个近大远小的场景，尽管你的眼睛或许会看到一种近乎平面状态的场景，但由于知觉层面观念性的作用，却也往往会固执地使你知觉到的是一种透视的场景。换言之，这是观念判断了对象物的场景本身。

近代，当人类能够乘热气球、飞机甚至如今乘宇宙飞船，从太空真正地俯瞰地球、俯瞰大地、俯瞰城市、俯瞰乡村、俯瞰大自然时，呈现在人类眼前的不再是基于透视所看到的近大远小，而是由于尺度上的巨大差别，一切都近乎扁平，一切都呈现出平面化的状态。原来的，一层又一层的，拥有层次的状态，被从顶部向下拍平。而如此这样的一种"拍平"的状态，带来的是将人们"透视"观念的直接"拍平"。于是，原有的拥有透视性的"写生"，继而便转换为平面化的，从空中和从顶部去观察到的世界。

这种从顶部向下看的"被拍平"的视角，反过来将其运用到平面中再去观察其中的一层一层的状态时，平

in which nearby objects seem large and distant objects small. Although your eyes may perceive almost two-dimensional scenes, you often obstinately stick to the perspective vision due to ingrained perception. In other words, perception makes its own judgment on scenes of objects themselves.

Recently by hot-air balloon, plane or even spaceship, when people can overlook the earth, land, cities, villages and nature from above, they see almost flat and two-dimensional scenes due to large disparity in size instead of the former perspective view. The former stratified status layer on layer is "beaten flat" from top to bottom. This kind of "beaten flat" status leads to the conception of "perspective" directly "beaten flat". Therefore, the former "paint from life" with perspective feature is consequently transformed into a two-dimensional world observed from the sky and from the top.

This viewing angle of "beaten flat" from top to bottom, when used in reverse to observe the stratified status contained within a two-dimensional plane, makes flat "patterns" on two-dimensional world themselves display stratified and overlaying sense. With this perception established, the sense of body and space, to some extent,

面中的平面化"图案"本身呈现出的是一个拥有层次和叠加感的世界。而伴随着这种知觉建立的同时，人的身体性与空间感，也在一定程度上发生着转变。

　　由此，这种从顶层向下的平面性的视点，最终，也使得写生这件事儿，从观念上得到转变。确切地说，过去的这种一层一层地对客观对象的描述，一旦进入从空中向下并呈现出被"拍平"的状态时，其实也就意味着观测"写生"对象物的视角发生了变化，意味着观察对象物的距离与尺度发生了变化，由此人对空间的理解自然而然地也同样在发生着变化。

　　这样的理解，事实上还正在带来传统意义上的所谓"写生"这个概念所要描绘的"世界"这个对象物的改变。一旦从这样的立场进行思考，那么当下学习美术的学生和建筑学的学生，不恰恰应该从日常的视点转化为从顶部、从空中、从宇宙的视点来对世界进行写生吗？

　　考虑到这里，《自然与建筑》这本书的宗旨便得以顺势浮出并呈现。我们这里所说的自然与建筑，是从上空俯瞰山川、树木与河流等地形地貌。而从空中去重新

changes consequently.

Therefore, this two-dimensional viewpoint from top to bottom also change paint from life finally in terms of conception. To be specific, the former layer-by-layer description about objective objects, when put into the status of top to bottom and "beaten flat", encounters changes in terms of not only viewpoint but also distance and scale to observe the objects being painted from life. Therefore people's understanding on space also changes naturally.

Such an understanding is actually changing the object of "world", the very object for the so-called conception of "paint from life" in traditional sense. Based on such a standpoint, then for students studying fine arts and architecture now, aren't they expected to paint the world from the top, from the sky and from the cosmos instead of adopting everyday viewpoint?

The above thoughts naturally give rise to the purpose of this book. *Nature and Architecture* as we mention in the book refers to a kind of observation on landforms such as mountains, trees and rivers from high above, which is a viewpoint we contemporary people are expected

观察我们周围的世界,这其实也是现代人所应该拓展的视角。对于建筑学的同仁而言,以这种来自"空中"的视点进行写生,是将视线放到一种更为广袤的地域范围中进行观察并获得的"世界"的写生。这种写生的根本性技法,就是用你的眼睛和视线,根据现场不同的区域与色彩,根据需要,去选择对象物本身的高低,用眼睛去使得对象物本身在意识中高起来、低下去。用眼睛调动意识并在意识空间中使对象物的"图案"本身的肌理呈现出"开敞"与"围合"。

当我们以这样的视点再去观察"世界"时,一切所谓我们以行走为基本行为的日常视点以及以行走在平面空间范围所看到的地域性,所看到的文化,都将会在观念中发生本质性的变化。而伴随着写生对象物的变化,平面的、扁平的世界与空间世界的连结关系便可以互相转换,并同时成为非"透视学"视点下的"写生"的一个主要特征。

to explore. For colleagues in architecture, to paint from life with this "aerial" viewpoint is to place sight line on a broader range to observe and acquire the world being painted. The fundamental skill of such a painting is to use your eyes and sight lines, according to different areas and colors on site as well as specific need, to choose the high and low of objects themselves, to use your eyes to make objects in your consciousness become higher or lower. Use your eyes to mobilize consciousness, and then make the texture of "patterns" reveal their "openness" and "closeness" in your consciousness space.

When the "world" is observed with such a viewpoint, our conception encounters fundamental changes in contrast to everyday viewpoint based on basic behavior of walking as well as regionalism and culture we perceive when walking on two-dimensional range. Together with the changes of objects for paint from life, there is interchangeability between flat two-dimensional world and three-dimensional world, which becomes a major feature of "paint from life" under non-perspective viewpoint.

前面我们对"写生"与"视点"的关系问题进行了探讨。本章节中，我们将对地形与建筑的关系进行描述与展示。在这里，我们从空间的视角对地形进行读解。首先，我们任意地从大自然中选择地形，在此基础上，我们对其进行图式化处理，并获得空间形态。在这个过程中发现其具有建筑价值与含义的空间形态是我们要考虑的目的。下面我们将以"空中"的视角对从大自然中所选取的19种自然地形肌理进行展示，并对其从地形转化为空间的过程进行逐一呈现。

We discussed relationship between "paint from life" and "viewpoint" before. In this chapter, we will describe and display the relationship between landform and architecture. Here we interpret landforms from a spatial perspective. Firstly, we randomly choose landforms from nature and then graphicalize these landforms to obtain spatial patterns. Our goal in this process is to discover spatial patterns containing architectural value and meaning. Next we will display 19 samples of natural landform textures observed from aerial perspective and reveal the transformation process from landform to space one by one.

从地形到空间的19个实例
19 samples of transformation from landform to space

实例2-1　从"北纬23.45'36.63",东经117.22'11.19""周边地形所选择出的建筑
Sample 2-1 Architecture chosen from landform around 23.45'36.63" N, 117.22'11.19" E

图2-1-1　"北纬23.45'36.63",东经117.22'11.19""周边地形卫星图
Figure 2-1-1 Satellite map of landform around 23.45'36.63" N, 117.22'11.19" E

图2-1-1所呈现的是从电子地图中截取的"北纬23.45'36.63″,东经117.22'11.19″"周边的地形。从这里,我们对所看到的拥有空间意义的部分进行抽取,并对其进行空间图式层面的转化(图2-1-2)。在此基础上,我们进一步对所获得的空间图式进行空间化处理,便可以获得拥有建筑空间意义的形态(图2-1-3)。图2-1-4和图2-1-5是所获得的空间形态在不同视角下的空间与形态的呈现。

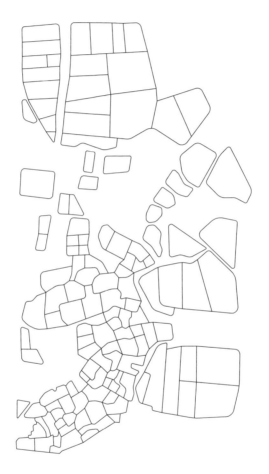

图2-1-2 从"北纬23.45'36.63″,东经117.22'11.19″"周边地形中所选取出的空间图式
Figure 2-1-2 Spatial diagrams chosen from landform around 23.45'36.63" N, 117.22'11.19" E

Figure 2-1-1 on the left page shows landform around 23.45'36.63" N, 117.22'11.19" E extracted from electronic maps. We extract parts with spatial meaning and conduct schematic transformation (figure 2-1-2). Based on this, we further conduct spatial processing on acquired spatial diagrams to obtain patterns with architectural spatial meaning (figure 2-1-3). Figure 2-1-4 and figure 2-1-5 represent space and pattern of acquired spatial patterns under different perspectives.

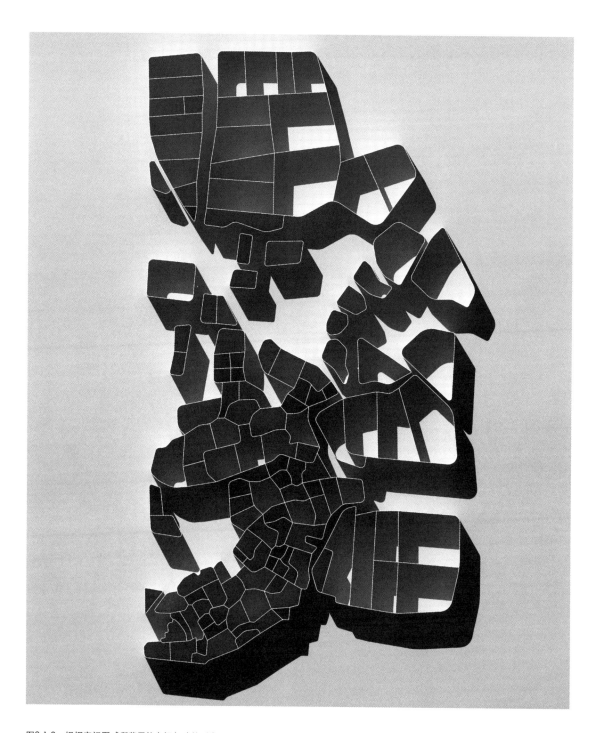

图2-1-3 根据空间图式所获得的空间与建筑形态
Figure 2-1-3 Spatial and architectural patterns acquired from spatial diagram

图2-1-4 空间与建筑形态在不同视角下的呈现
Figure 2-1-4 Space and architectural patterns under different perspectives

图2-1-5 空间与建筑形态在不同视角下的呈现
Figure 2-1-5 Space and architectural patterns under different perspectives

实例2-2 从"北纬23.54'55.90",东经57.06'02.58""周边地形所选择出的建筑
Sample 2-2 Architecture chosen from landform around 23.54'55.90" N, 57.06'02.58" E

图2-2-1 "北纬23.54'55.90",东经57.06'02.58""周边地形卫星图
Figure 2-2-1 Satellite map of landform around 23.54'55.90" N, 57.06'02.58" E

图2-2-1所呈现的是从电子地图中截取的 "北纬23.54'55.90″, 东经57.06'02.58″" 周边的地形。从这里, 我们对所看到的拥有空间意义的部分进行抽取, 并对其进行空间图式层面的转化（图2-2-2）。在此基础上, 我们进一步对所获得的空间图式进行空间化处理, 便可以获得拥有建筑空间意义的形态（图2-2-3）。图2-2-4是所获得的空间形态在不同视角下的空间与形态的呈现。

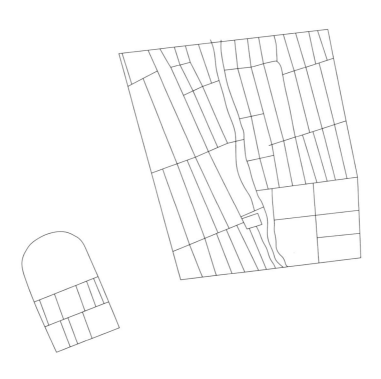

图2-2-2 从 "北纬23.54'55.90″, 东经57.06'02.58″" 周边地形中所选取出的空间图式
Figure 2-2-2 Spatial diagrams chosen from landform around 23.54'55.90" N, 57.06'02.58" E

Figure 2-2-1 on the left page shows landform around 23.54'55.90" N, 57.06'02.58" E extracted from electronic maps. We extract parts with spatial meaning and conduct schematic transformation (figure 2-2-2). Based on this, we further conduct spatial processing on acquired spatial diagrams to obtain patterns with architectural spatial meaning (figure 2-2-3). Figure 2-2-4 represents space and pattern of acquired spatial patterns under different perspectives.

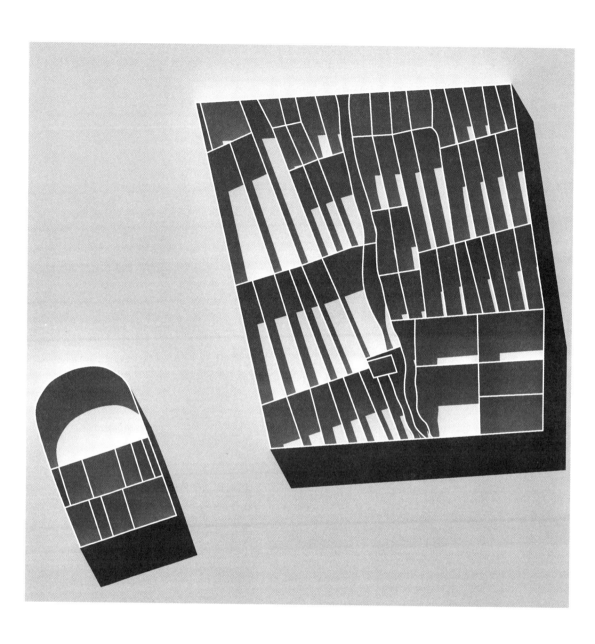

图2-2-3 根据空间图式所获得的空间与建筑形态
Figure 2-2-3 Spatial and architectural patterns acquired from spatial diagram

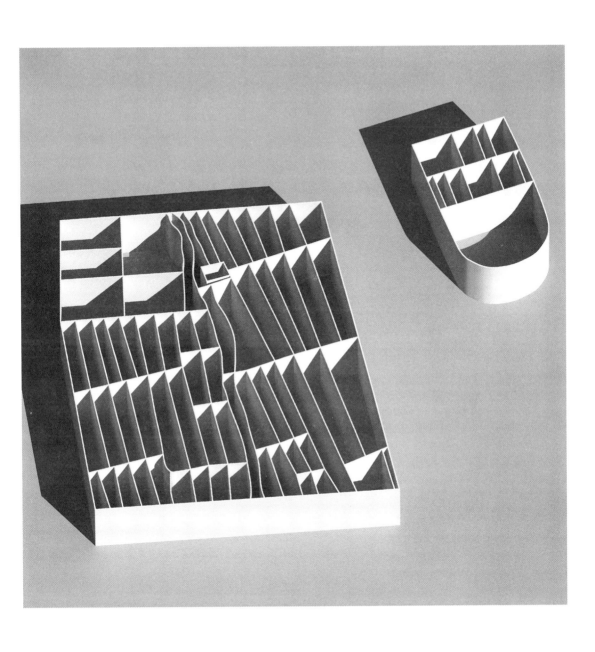

图2-2-4 空间与建筑形态在不同视角下的呈现
Figure 2-2-4 Space and architectural patterns under different perspectives

实例2-3 从"北纬31.52'20.67″,东经36.58'39.98″"周边地形所选择出的建筑
Sample 2-3 Architecture chosen from landform around 31.52'20.67" N, 36.58'39.98" E

图2-3-1 "北纬31.52'20.67″,东经36.58'39.98″"周边地形卫星图
Figure 2-3-1 Satellite map of landform around 31.52'20.67" N, 36.58'39.98" E

图2-3-1所呈现的是从电子地图中截取的"北纬31.52'20.67",东经36.58'39.98""周边的地形。从这里,我们对所看到的拥有空间意义的部分进行抽取,并对其进行空间图式层面的转化(图2-3-2)。在此基础上,我们进一步对所获得的空间图式进行空间化处理,便可以获得拥有建筑空间意义的形态(图2-3-3)。图2-3-4和图2-3-5是所获得的空间形态在不同视角下的空间与形态的呈现。

图2-3-2　从"北纬31.52'20.67",东经36.58'39.98""周边地形中所选取出的空间图式
Figure 2-3-2 Spatial diagrams chosen from landform around 31.52'20.67" N, 36.58'39.98" E

Figure 2-3-1 on the left page shows landform around 31.52'20.67" N, 36.58'39.98" E extracted from electronic maps. We extract parts with spatial meaning and conduct schematic transformation (figure 2-3-2). Based on this, we further conduct spatial processing on acquired spatial diagrams to obtain patterns with architectural spatial meaning (figure 2-3-3). Figure 2-3-4 and figure 2-3-5 represent space and pattern of acquired spatial patterns under different perspectives.

图2-3-3 根据空间图式所获得的空间与建筑形态
Figure 2-3-3 Spatial and architectural patterns acquired from spatial diagram

图2-3-4 空间与建筑形态在不同视角下的呈现
Figure 2-3-4 Space and architectural patterns under different perspectives

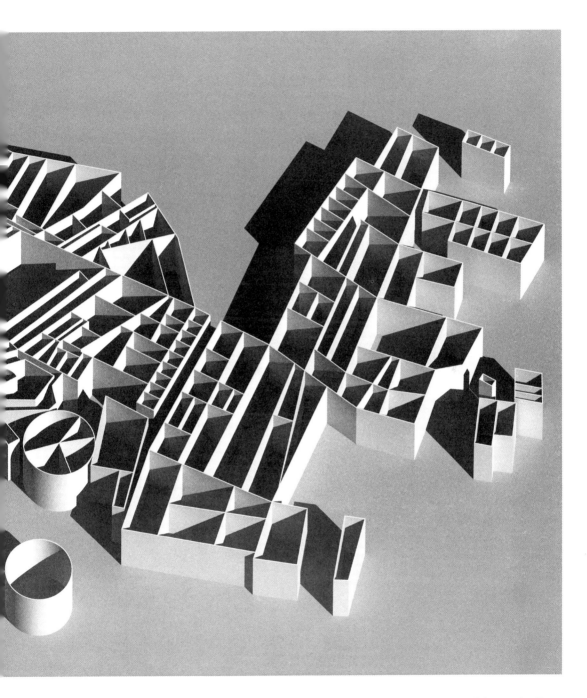

图2-3-5 空间与建筑形态在不同视角下的呈现
Figure 2-3-5 Space and architectural patterns under different perspectives

实例2-4 从"北纬31.53'02.68″,东经36.52'23.09″"周边地形所选择出的建筑
Sample 2-4 Architecture chosen from landform around 31.53'02.68" N, 36.52'23.09" E

图2-4-1 "北纬31.53'02.68″,东经36.52'23.09″"周边地形卫星图
Figure 2-4-1 Satellite map of landform around 31.53'02.68" N, 36.52'23.09" E

图2-4-1所呈现的是从电子地图中截取的"北纬31.53'02.68",东经36.52'23.09""周边的地形。从这里,我们对所看到的拥有空间意义的部分进行抽取,并对其进行空间图式层面的转化(图2-4-2)。在此基础上,我们进一步对所获得的空间图式进行空间化处理,便可以获得拥有建筑空间意义的形态(图2-4-3)。图2-4-4和图2-4-5是所获得的空间形态在不同视角下的空间与形态的呈现。

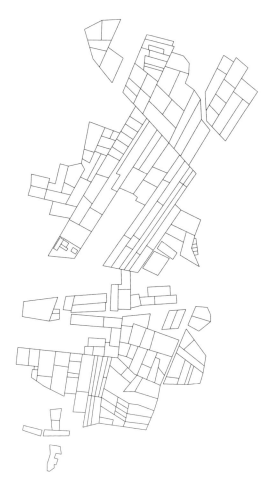

图2-4-2 从"北纬31.53'02.68",东经36.52'23.09""周边地形中所选取出的空间图式
Figure 2-4-2 Spatial diagrams chosen from landform around 31.53'02.68" N, 36.52'23.09" E

Figure 2-4-1 on the left page shows landform around 31.53'02.68" N, 36.52'23.09" E extracted from electronic maps. We extract parts with spatial meaning and conduct schematic transformation (figure 2-4-2). Based on this, we further conduct spatial processing on acquired spatial diagrams to obtain patterns with architectural spatial meaning (figure 2-4-3). Figure 2-4-4 and figure 2-4-5 represent space and pattern of acquired spatial patterns under different perspectives.

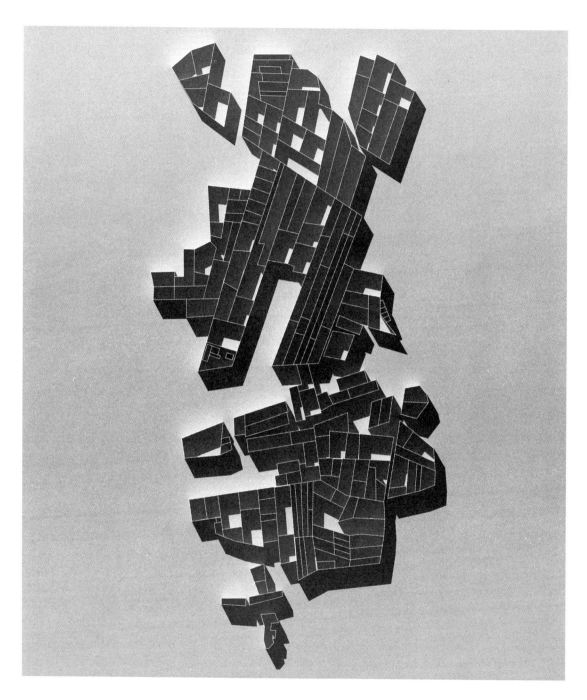

图2-4-3 根据空间图式所获得的空间与建筑形态

Figure 2-4-3 Spatial and architectural patterns acquired from spatial diagram

图2-4-4 空间与建筑形态在不同视角下的呈现
Figure 2-4-4 Space and architectural patterns under different perspectives

图2-4-5 空间与建筑形态在不同视角下的呈现
Figure 2-4-5 Space and architectural patterns under different perspectives

实例2-5 从 "北纬32.11'20.17"，东经36.08'46.05""周边地形所选择出的建筑
Sample 2-5 Architecture chosen from landform around 32.11'20.17" N, 36.08'46.05" E

图2-5-1 "北纬32.11'20.17"，东经36.08'46.05""周边地形卫星图
Figure 2-5-1 Satellite map of landform around 32.11'20.17" N, 36.08'46.05" E

图2-5-1所呈现的是从电子地图中截取的"北纬32.11'20.17″,东经36.08'46.05″"周边的地形。从这里,我们对所看到的拥有空间意义的部分进行抽取,并对其进行空间图式层面的转化(图2-5-2)。在此基础上,我们进一步对所获得的空间图式进行空间化处理,便可以获得拥有建筑空间意义的形态(图2-5-3)。图2-5-4和图2-5-5是所获得的空间形态在不同视角下的空间与形态的呈现。

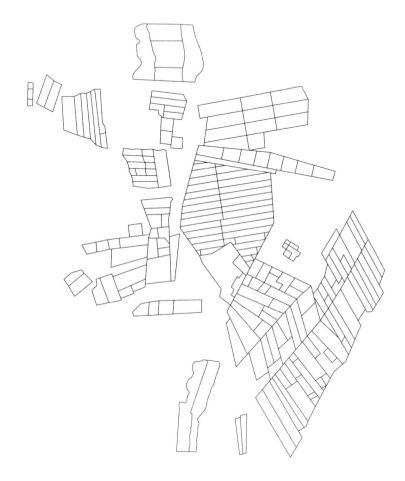

图2-5-2 从"北纬32.11'20.17″,东经36.08'46.05″"周边地形中所选取出的空间图式
Figure 2-5-2 Spatial diagrams chosen from landform around 32.11'20.17" N, 36.08'46.05" E

Figure 2-5-1 on the left page shows landform around 32.11'20.17" N, 36.08'46.05" E extracted from electronic maps. We extract parts with spatial meaning and conduct schematic transformation (figure 2-5-2). Based on this, we further conduct spatial processing on acquired spatial diagrams to obtain patterns with architectural spatial meaning (figure 2-5-3). Figure 2-5-4 and figure 2-5-5 represent space and pattern of acquired spatial patterns under different perspectives.

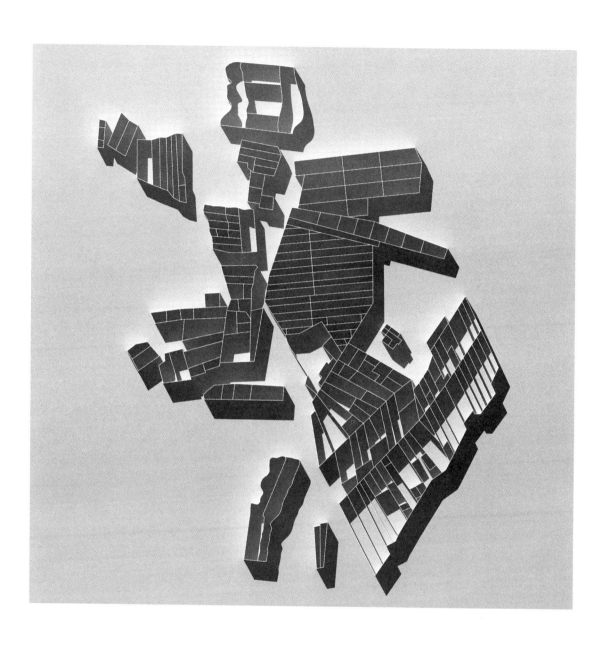

图2-5-3 根据空间图式所获得的空间与建筑形态

Figure 2-5-3 Spatial and architectural patterns acquired from spatial diagram

图2-5-4 空间与建筑形态在不同视角下的呈现
Figure 2-5-4 Space and architectural patterns under different perspectives

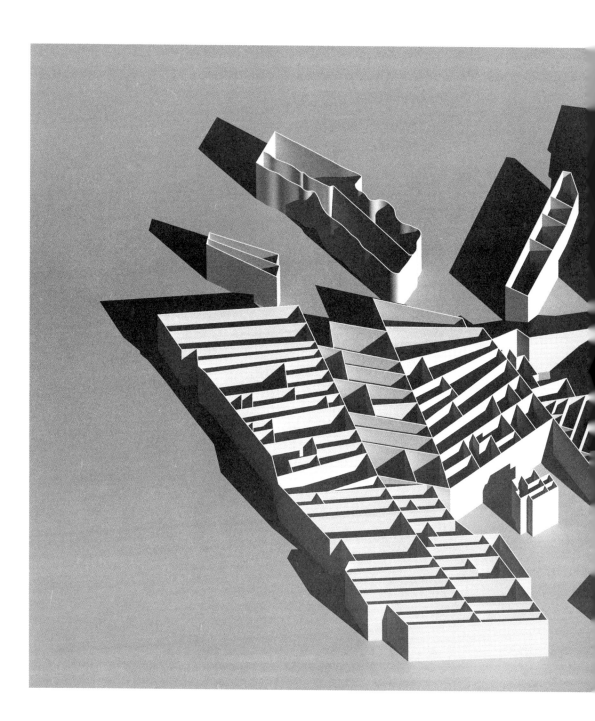

图2-5-5 空间与建筑形态在不同视角下的呈现
Figure 2-5-5 Space and architectural patterns under different perspectives

实例2-6 从"北纬23.45'01.54",东经117.14'03.03""周边地形所选择出的建筑
Sample 2-6 Architecture chosen from landform around 23.45'01.54" N, 117.14'03.03" E

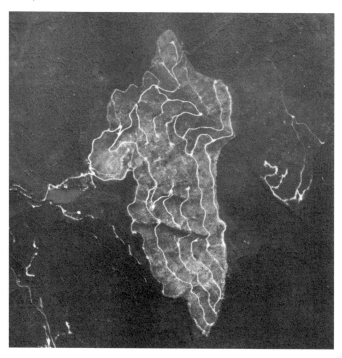

图2-6-1 "北纬23.45'01.54",东经117.14'03.03""周边地形卫星图
Figure 2-6-1 Satellite map of landform around 23.45'01.54" N, 117.14'03.03" E

图2-6-1所呈现的是从电子地图中截取的"北纬23.45'01.54″,东经117.14'03.03″"周边的地形。从这里,我们对所看到的拥有空间意义的部分进行抽取,并对其进行空间图式层面的转化(图2-6-2)。在此基础上,我们进一步对所获得的空间图式进行空间化处理,便可以获得拥有建筑空间意义的形态(图2-6-3)。图2-6-4至图2-6-6是所获得的空间形态在不同视角下的空间与形态的呈现。

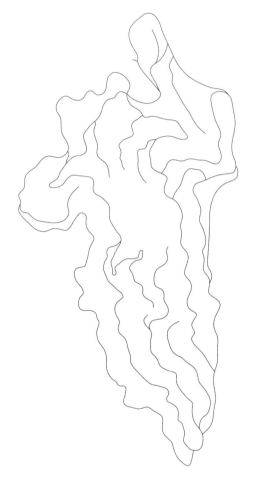

图2-6-2 从"北纬23.45'01.54″,东经117.14'03.03″"周边地形中所选取出的空间图式
Figure 2-6-2 Spatial diagrams chosen from landform around 23.45'01.54" N, 117.14'03.03" E

Figure 2-6-1 on the left page shows landform around 23.45'01.54" N, 117.14'03.03" E extracted from electronic maps. We extract parts with spatial meaning and conduct schematic transformation (figure 2-6-2). Based on this, we further conduct spatial processing on acquired spatial diagrams to obtain patterns with architectural spatial meaning (figure 2-6-3). Figure 2-6-4 to figure 2-6-6 represent space and pattern of acquired spatial patterns under different perspectives.

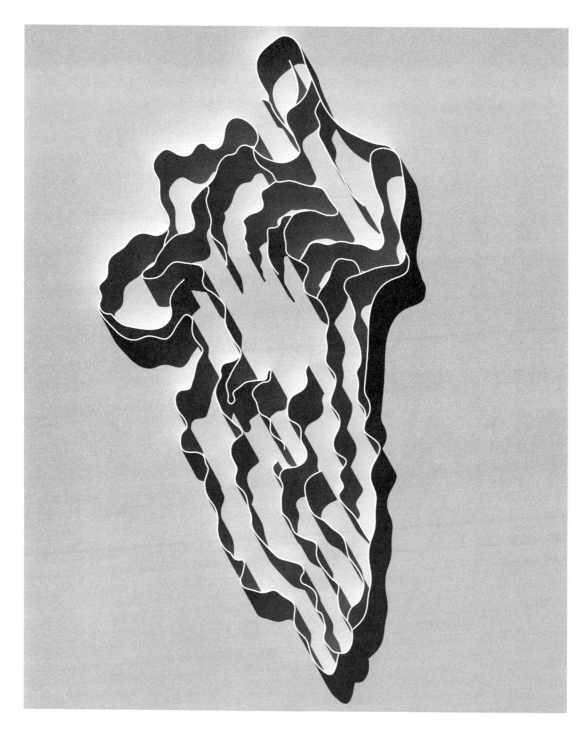

图2-6-3 根据空间图式所获得的空间与建筑形态

Figure 2-6-3 Spatial and architectural patterns acquired from spatial diagram

图2-6-4 空间与建筑形态在不同视角下的呈现
Figure 2-6-4 Space and architectural patterns under different perspectives

图2-6-5 空间与建筑形态在不同视角下的呈现
Figure 2-6-5 Space and architectural patterns under different perspectives

图2-6-6 空间与建筑形态在不同视角下的呈现
Figure 2-6-6 Space and architectural patterns under different perspectives

实例2-7 从 "北纬32.13'03.67″,东经36.08'35.01″" 周边地形所选择出的建筑
Sample 2-7 Architecture chosen from landform around 32.13'03.67" N, 36.08'35.01" E

图2-7-1 "北纬32.13'03.67″,东经36.08'35.01″" 周边地形卫星图
Figure 2-7-1 Satellite map of landform around 32.13'03.67" N, 36.08'35.01" E

图2-7-1所呈现的是从电子地图中截取的 "北纬32.13'03.67", 东经36.08'35.01"" 周边的地形。从这里, 我们对所看到的拥有空间意义的部分进行抽取, 并对其进行空间图式层面的转化 (图2-7-2) 。在此基础上, 我们进一步对所获得的空间图式进行空间化处理, 便可以获得拥有建筑空间意义的形态 (图2-7-3) 。图2-7-4至图2-7-6是所获得的空间形态在不同视角下的空间与形态的呈现。

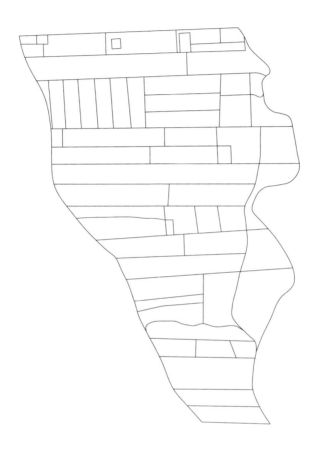

图2-7-2 从 "北纬32.13'03.67", 东经36.08'35.01"" 周边地形中所选取出的空间图式
Figure 2-7-2 Spatial diagrams chosen from landform around 32.13'03.67" N, 36.08'35.01" E

Figure 2-7-1 on the left page shows landform around 32.13'03.67" N, 36.08'35.01" E extracted from electronic maps. We extract parts with spatial meaning and conduct schematic transformation (figure 2-7-2). Based on this, we further conduct spatial processing on acquired spatial diagrams to obtain patterns with architectural spatial meaning (figure 2-7-3). Figure 2-7-4 to figure 2-7-6 represent space and pattern of acquired spatial patterns under different perspectives.

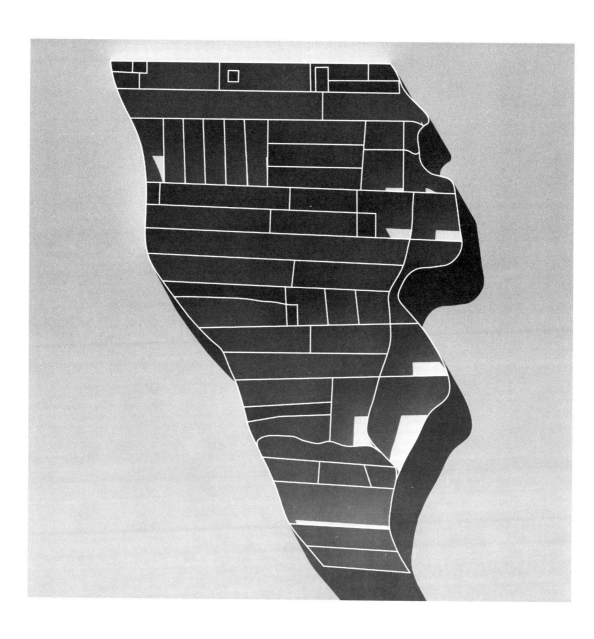

图2-7-3 根据空间图式所获得的空间与建筑形态

Figure 2-7-3 Spatial and architectural patterns acquired from spatial diagram

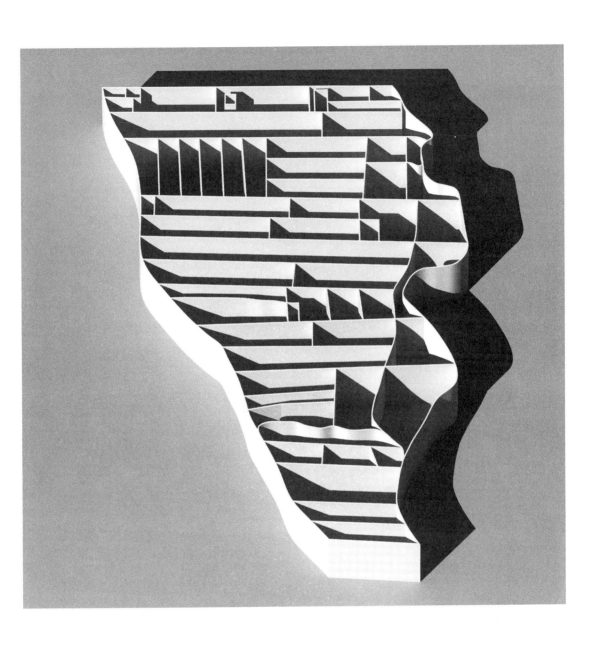

图2-7-4 空间与建筑形态在不同视角下的呈现
Figure 2-7-4 Space and architectural patterns under different perspectives

图2-7-5 空间与建筑形态在不同视角下的呈现
Figure 2-7-5 Space and architectural patterns under different perspectives

图2-7-6 空间与建筑形态在不同视角下的呈现
Figure 2-7-6 Space and architectural patterns under different perspectives

实例2-8 从"北纬32.13'10.41",东经36.14'03.79""周边地形所选择出的建筑
Sample 2-8 Architecture chosen from landform around 32.13'10.41" N, 36.14'03.79" E

图2-8-1 "北纬32.13'10.41",东经36.14'03.79""周边地形卫星图
Figure 2-8-1 Satellite map of landform around 32.13'10.41" N, 36.14'03.79" E

图2-8-1所呈现的是从电子地图中截取的"北纬32.13'10.41",东经36.14'03.79""周边的地形。从这里,我们对所看到的拥有空间意义的部分进行抽取,并对其进行空间图式层面的转化(图2-8-2)。在此基础上,我们进一步对所获得的空间图式进行空间化处理,便可以获得拥有建筑空间意义的形态(图2-8-3)。图2-8-4至图2-8-6是所获得的空间形态在不同视角下的空间与形态的呈现。

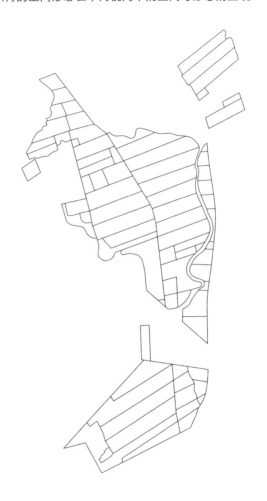

图2-8-2 从"北纬32.13'10.41",东经36.14'03.79""周边地形中所选取出的空间图式
Figure 2-8-2 Spatial diagrams chosen from landform around 32.13'10.41" N, 36.14'03.79" E

Figure 2-8-1 on the left page shows landform around 32.13'10.41" N, 36.14'03.79" E extracted from electronic maps. We extract parts with spatial meaning and conduct schematic transformation (figure 2-8-2). Based on this, we further conduct spatial processing on acquired spatial diagrams to obtain patterns with architectural spatial meaning (figure 2-8-3). Figure 2-8-4 to figure 2-8-6 represent space and pattern of acquired spatial patterns under different perspectives.

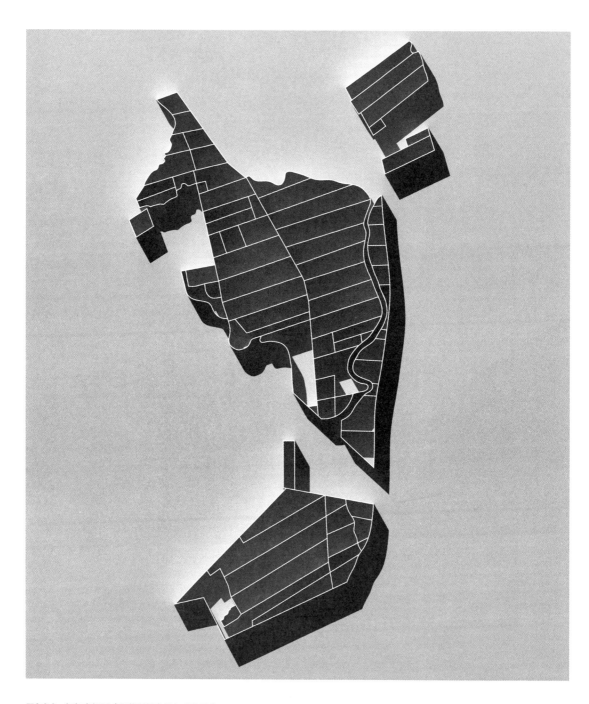

图2-8-3 根据空间图式所获得的空间与建筑形态

Figure 2-8-3 Spatial and architectural patterns acquired from spatial diagram

图2-8-4 空间与建筑形态在不同视角下的呈现
Figure 2-8-4 Space and architectural patterns under different perspectives

图2-8-5 空间与建筑形态在不同视角下的呈现
Figure 2-8-5 Space and architectural patterns under different perspectives

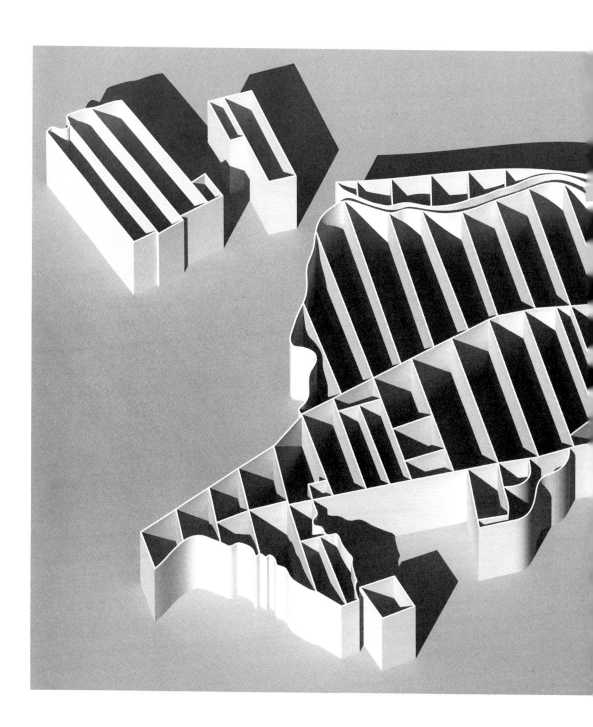

图2-8-6 空间与建筑形态在不同视角下的呈现
Figure 2-8-6 Space and architectural patterns under different perspectives

实例2-9 从"北纬23.54'27.80",东经117.28'36.57""周边地形所选择出的建筑
Sample 2-9 Architecture chosen from landform around 23.54'27.80" N, 117.28'36.57" E

图2-9-1 "北纬23.54'27.80",东经117.28'36.57""周边地形卫星图
Figure 2-9-1 Satellite map of landform around 23.54'27.80" N, 117.28'36.57" E

图2-9-1所呈现的是从电子地图中截取的"北纬23.54'27.80'',东经117.28'36.57''"周边的地形。从这里,我们对所看到的拥有空间意义的部分进行抽取,并对其进行空间图式层面的转化(图2-9-2)。在此基础上,我们进一步对所获得的空间图式进行空间化处理,便可以获得拥有建筑空间意义的形态(图2-9-3)。图2-9-4和图2-9-5是所获得的空间形态在不同视角下的空间与形态的呈现。

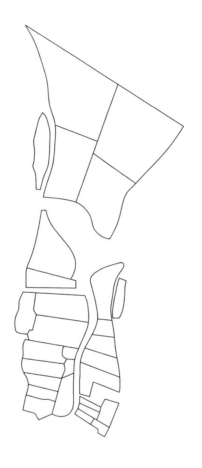

图2-9-2 从"北纬23.54'27.80'',东经117.28'36.57''"周边地形中所选取出的空间图式
Figure 2-9-2 Spatial diagrams chosen from landform around 23.54'27.80" N, 117.28'36.57" E

Figure 2-9-1 on the left page shows landform around 23.54'27.80" N, 117.28'36.57" E extracted from electronic maps. We extract parts with spatial meaning and conduct schematic transformation (figure 2-9-2). Based on this, we further conduct spatial processing on acquired spatial diagrams to obtain patterns with architectural spatial meaning (figure 2-9-3). Figure 2-9-4 and figure 2-9-5 represent space and pattern of acquired spatial patterns under different perspectives.

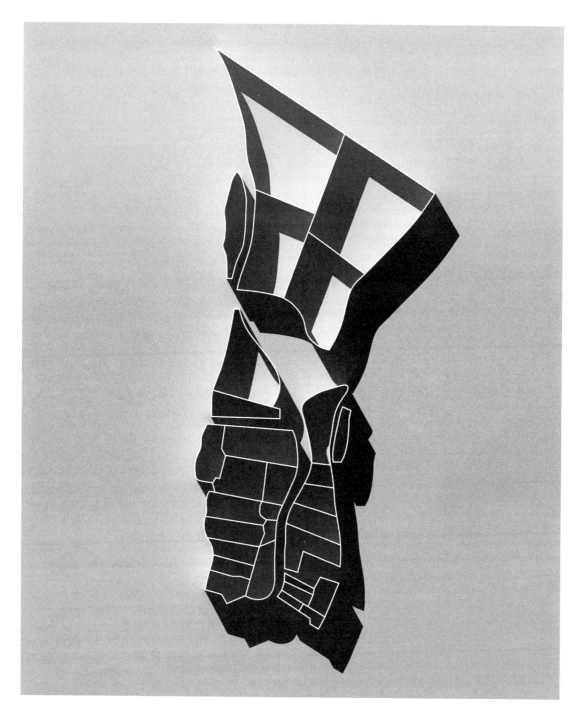

图2-9-3 根据空间图式所获得的空间与建筑形态

Figure 2-9-3 Spatial and architectural patterns acquired from spatial diagram

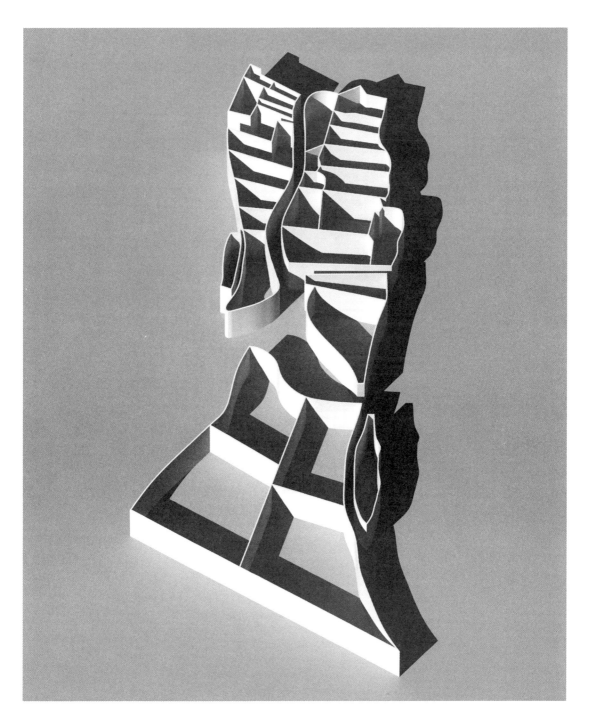

图2-9-4 空间与建筑形态在不同视角下的呈现
Figure 2-9-4 Space and architectural patterns under different perspectives

图2-9-5 空间与建筑形态在不同视角下的呈现
Figure 2-9-5 Space and architectural patterns under different perspectives

实例2-10 从"北纬32.06'07.90",东经37.22'11.49""周边地形所选择出的建筑
Sample 2-10 Architecture chosen from landform around 32.06'07.90" N, 37.22'11.49" E

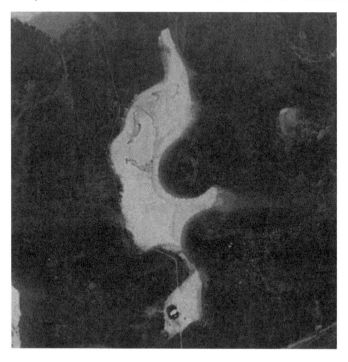

图2-10-1 "北纬32.06'07.90",东经37.22'11.49""周边地形卫星图
Figure 2-10-1 Satellite map of landform around 32.06'07.90" N, 37.22'11.49" E

图2-10-1所呈现的是从电子地图中截取的"北纬32.06'07.90″,东经37.22'11.49″"周边的地形。从这里,我们对所看到的拥有空间意义的部分进行抽取,并对其进行空间图式层面的转化(图2-10-2)。在此基础上,我们进一步对所获得的空间图式进行空间化处理,便可以获得拥有建筑空间意义的形态(图2-10-3)。图2-10-4至图2-10-6是所获得的空间形态在不同视角下的空间与形态的呈现。

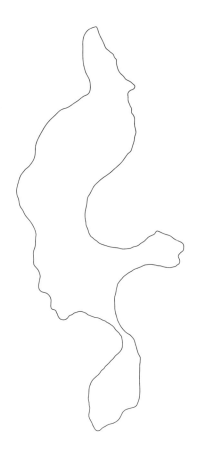

图2-10-2 从"北纬32.06'07.90″,东经37.22'11.49″"周边地形中所选取出的空间图式
Figure 2-10-2 Spatial diagrams chosen from landform around 32.06'07.90" N, 37.22'11.49" E

Figure 2-10-1 on the left page shows landform around 32.06'07.90" N, 37.22'11.49" E extracted from electronic maps. We extract parts with spatial meaning and conduct schematic transformation (figure 2-10-2). Based on this, we further conduct spatial processing on acquired spatial diagrams to obtain patterns with architectural spatial meaning (figure 2-10-3). Figure 2-10-4 to figure 2-10-6 represent space and pattern of acquired spatial patterns under different perspectives.

图2-10-3 根据空间图式所获得的空间与建筑形态
Figure 2-10-3 Spatial and architectural patterns acquired from spatial diagram

图2-10-4 空间与建筑形态在不同视角下的呈现
Figure 2-10-4 Space and architectural patterns under different perspectives

图2-10-5 空间与建筑形态在不同视角下的呈现
Figure 2-10-5 Space and architectural patterns under different perspectives

图2-10-6 空间与建筑形态在不同视角下的呈现
Figure 2-10-6 Space and architectural patterns under different perspectives

实例2-11 从"北纬14.56'29.75″,东经43.23'25.95″"周边地形所选择出的建筑
Sample 2-11 Architecture chosen from landform around 14.56'29.75" N, 43.23'25.95" E

图2-11-1 "北纬14.56'29.75″,东经43.23'25.95″"周边地形卫星图
Figure 2-11-1 Satellite map of landform around 14.56'29.75" N, 43.23'25.95" E

图2-11-1所呈现的是从电子地图中截取的"北纬14.56′29.75″,东经43.23′25.95″"周边的地形。从这里,我们对所看到的拥有空间意义的部分进行抽取,并对其进行空间图式层面的转化(图2-11-2)。在此基础上,我们进一步对所获得的空间图式进行空间化处理,便可以获得拥有建筑空间意义的形态(图2-11-3)。图2-11-4和图2-11-5是所获得的空间形态在不同视角下的空间与形态的呈现。

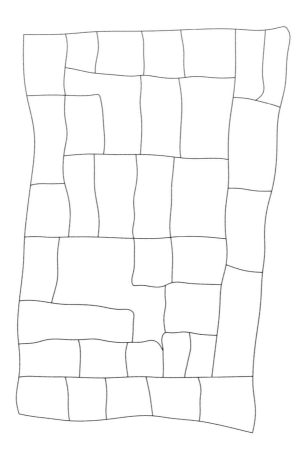

图2-11-2 从"北纬14.56′29.75″,东经43.23′25.95″"周边地形中所选取出的空间图式
Figure 2-11-2 Spatial diagrams chosen from landform around 14.56′29.75″ N, 43.23′25.95″ E

Figure 2-11-1 on the left page shows landform around 14.56′29.75″ N, 43.23′25.95″ E extracted from electronic maps. We extract parts with spatial meaning and conduct schematic transformation (figure 2-11-2). Based on this, we further conduct spatial processing on acquired spatial diagrams to obtain patterns with architectural spatial meaning (figure 2-11-3). Figure 2-11-4 and figure 2-11-5 represent space and pattern of acquired spatial patterns under different perspectives.

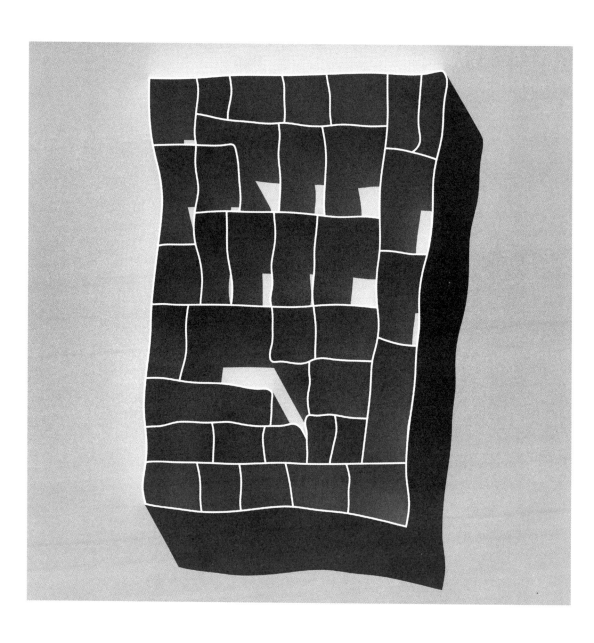

图2-11-3 根据空间图式所获得的空间与建筑形态
Figure 2-11-3 Spatial and architectural patterns acquired from spatial diagram

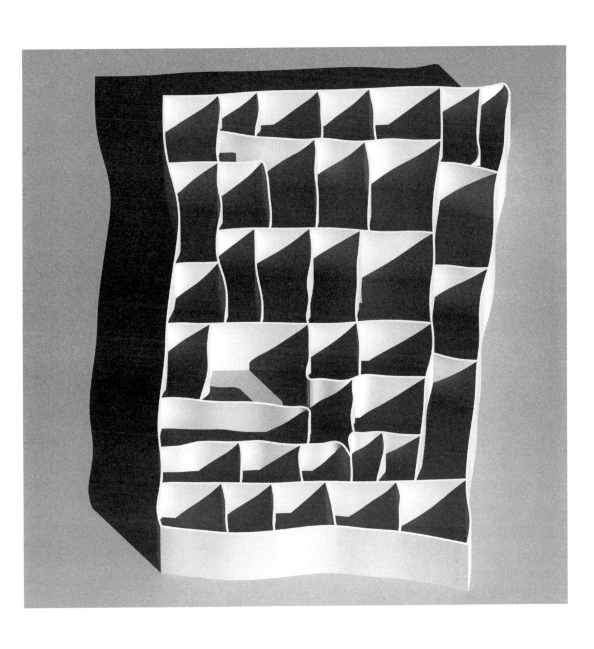

图2-11-4 空间与建筑形态在不同视角下的呈现
Figure 2-11-4 Space and architectural patterns under different perspectives

图2-11-5 空间与建筑形态在不同视角下的呈现
Figure 2-11-5 Space and architectural patterns under different perspectives

实例2-12 从"北纬14.52'56.87",东经43.17'50.12""周边地形所选择出的建筑
Sample 2-12 Architecture chosen from landform around 14.52'56.87" N, 43.17'50.12" E

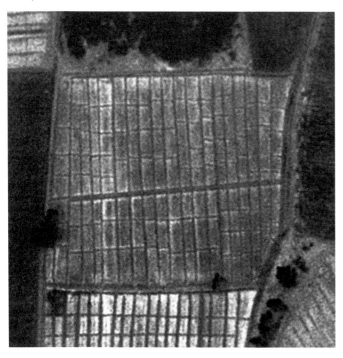

图2-12-1 "北纬14.52'56.87",东经43.17'50.12""周边地形卫星图
Figure 2-12-1 Satellite map of landform around 14.52'56.87" N, 43.17'50.12" E

图2-12-1所呈现的是从电子地图中截取的"北纬14.52'56.87",东经43.17'50.12""周边的地形。从这里,我们对所看到的拥有空间意义的部分进行抽取,并对其进行空间图式层面的转化(图2-12-2)。在此基础上,我们进一步对所获得的空间图式进行空间化处理,便可以获得拥有建筑空间意义的形态(图2-12-3)。图2-12-4是所获得的空间形态在不同视角下的空间与形态的呈现。

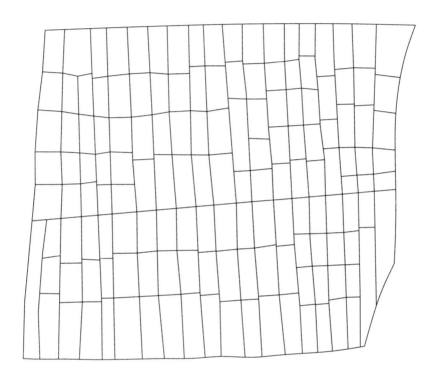

图2-12-2 从"北纬14.52'56.87",东经43.17'50.12""周边地形中所选取出的空间图式
Figure 2-12-2 Spatial diagrams chosen from landform around 14.52'56.87" N, 43.17'50.12" E

Figure 2-12-1 on the left page shows landform around 14.52'56.87" N, 43.17'50.12" E extracted from electronic maps. We extract parts with spatial meaning and conduct schematic transformation (figure 2-12-2). Based on this, we further conduct spatial processing on acquired spatial diagrams to obtain patterns with architectural spatial meaning (figure 2-12-3). Figure 2-12-4 represents space and pattern of acquired spatial patterns under different perspectives.

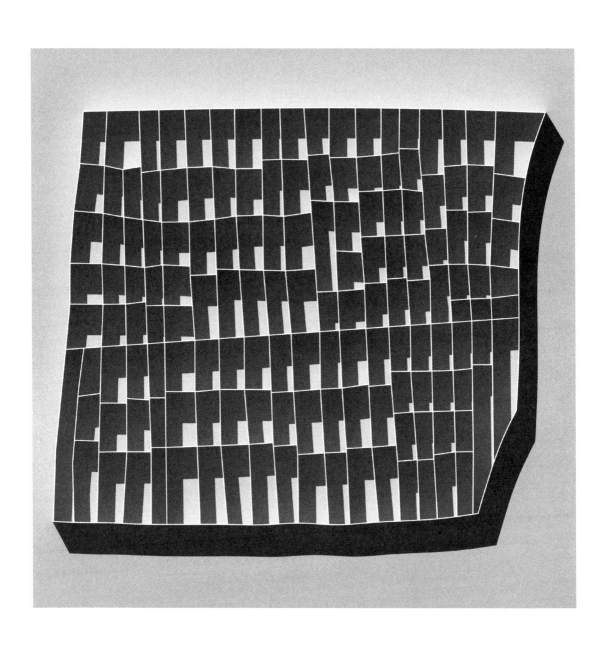

图2-12-3 根据空间图式所获得的空间与建筑形态
Figure 2-12-3 Spatial and architectural patterns acquired from spatial diagram

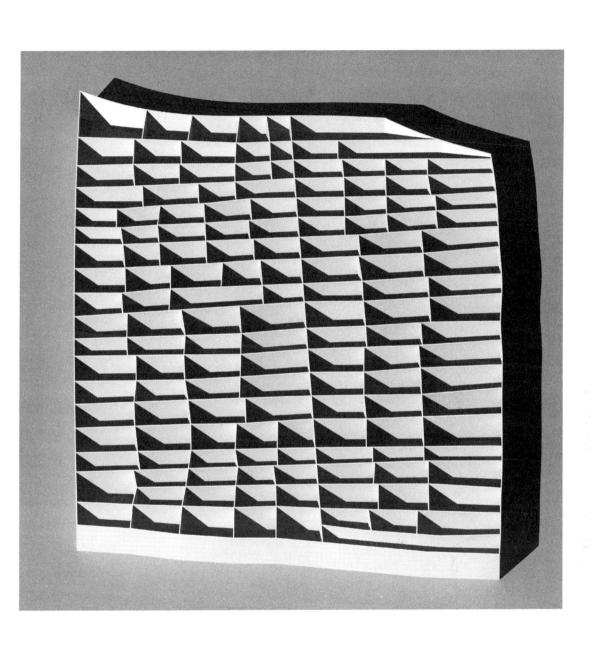

图2-12-4 空间与建筑形态在不同视角下的呈现
Figure 2-12-4 Space and architectural patterns under different perspectives

实例2-13 从"北纬32.16'08.40",东经36.05'48.26""周边地形所选择出的建筑
Sample 2-13 Architecture chosen from landform around 32.16'08.40" N, 36.05'48.26" E

图2-13-1 "北纬32.16'08.40",东经36.05'48.26""周边地形卫星图
Figure 2-13-1 Satellite map of landform around 32.16'08.40" N, 36.05'48.26" E

图2-13-1所呈现的是从电子地图中截取的"北纬32.16'08.40",东经36.05'48.26""周边的地形。从这里,我们对所看到的拥有空间意义的部分进行抽取,并对其进行空间图式层面的转化(图2-13-2)。在此基础上,我们进一步对所获得的空间图式进行空间化处理,便可以获得拥有建筑空间意义的形态(图2-13-3)。图2-13-4和图2-13-5是所获得的空间形态在不同视角下的空间与形态的呈现。

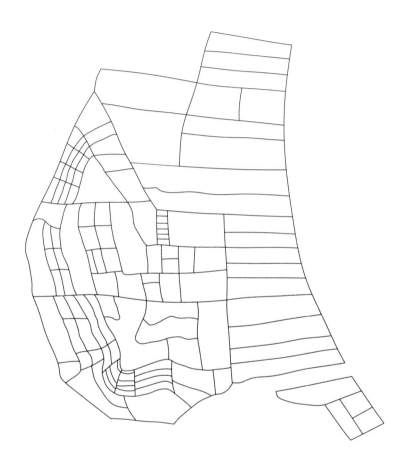

图2-13-2 从"北纬32.16'08.40",东经36.05'48.26""周边地形中所选取出的空间图式
Figure 2-13-2 Spatial diagrams chosen from landform around 32.16'08.40" N, 36.05'48.26" E

Figure 2-13-1 on the left page shows landform around 32.16'08.40" N, 36.05'48.26" E extracted from electronic maps. We extract parts with spatial meaning and conduct schematic transformation (figure 2-13-2). Based on this, we further conduct spatial processing on acquired spatial diagrams to obtain patterns with architectural spatial meaning (figure 2-13-3). Figure 2-13-4 and figure 2-13-5 represent space and pattern of acquired spatial patterns under different perspectives.

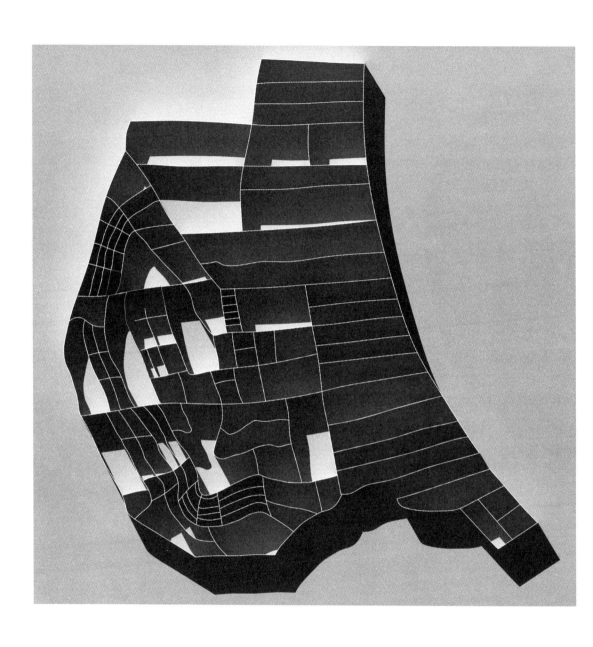

图2-13-3 根据空间图式所获得的空间与建筑形态
Figure 2-13-3 Spatial and architectural patterns acquired from spatial diagram

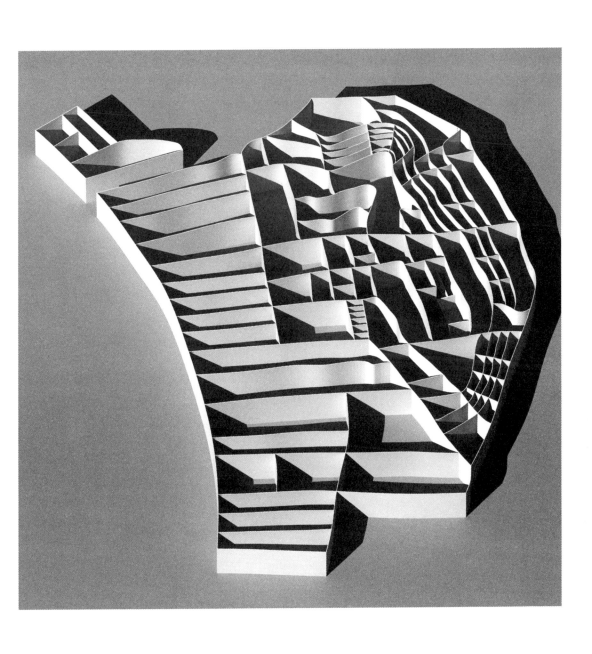

图2-13-4 空间与建筑形态在不同视角下的呈现
Figure 2-13-4 Space and architectural patterns under different perspectives

图2-13-5 空间与建筑形态在不同视角下的呈现
Figure 2-13-5 Space and architectural patterns under different perspectives

实例2-14 从"北纬53.50'51.13",东经73.24'12.32""周边地形所选择出的建筑
Sample 2-14 Architecture chosen from landform around 53.50'51.13" N, 73.24'12.32" E

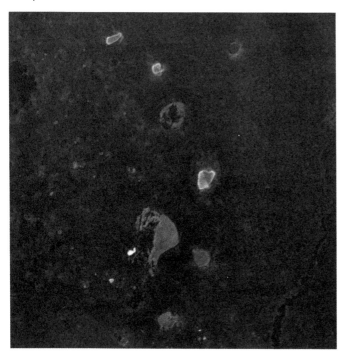

图2-14-1 "北纬53.50'51.13",东经73.24'12.32""周边地形卫星图
Figure 2-14-1 Satellite map of landform around 53.50'51.13" N, 73.24'12.32" E

图2-14-1所呈现的是从电子地图中截取的"北纬53.50'51.13″,东经73.24'12.32″"周边的地形。从这里,我们对所看到的拥有空间意义的部分进行抽取,并对其进行空间图式层面的转化(图2-14-2)。在此基础上,我们进一步对所获得的空间图式进行空间化处理,便可以获得拥有建筑空间意义的形态(图2-14-3)。图2-14-4和图2-14-5是所获得的空间形态在不同视角下的空间与形态的呈现。

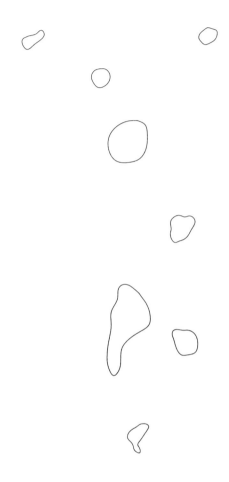

图2-14-2 从"北纬53.50'51.13″,东经73.24'12.32″"周边地形中所选取出的空间图式
Figure 2-14-2 Spatial diagrams chosen from landform around 53.50'51.13" N, 73.24'12.32" E

Figure 2-14-1 on the left page shows landform around 53.50'51.13" N, 73.24'12.32" E extracted from electronic maps. We extract parts with spatial meaning and conduct schematic transformation (figure 2-14-2). Based on this, we further conduct spatial processing on acquired spatial diagrams to obtain patterns with architectural spatial meaning (figure 2-14-3). Figure 2-14-4 and figure 2-14-5 represent space and pattern of acquired spatial patterns under different perspectives.

图2-14-3 根据空间图式所获得的空间与建筑形态

Figure 2-14-3 Spatial and architectural patterns acquired from spatial diagram

图2-14-4 空间与建筑形态在不同视角下的呈现
Figure 2-14-4 Space and architectural patterns under different perspectives

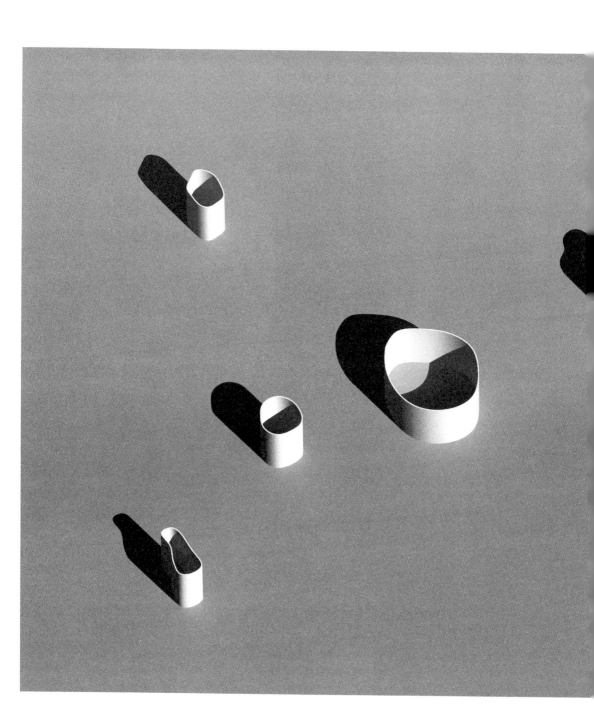

图2-14-5 空间与建筑形态在不同视角下的呈现
Figure 2-14-5 Space and architectural patterns under different perspectives

实例2-15 从"北纬34.17'24.41",东经57.33'20.18""周边地形所选择出的建筑

Sample 2-15 Architecture chosen from landform around 34.17'24.41" N, 57.33'20.18" E

图2-15-1 "北纬34.17'24.41",东经57.33'20.18""周边地形卫星图

Figure 2-15-1 Satellite map of landform around 34.17'24.41" N, 57.33'20.18" E

图2-15-1所呈现的是从电子地图中截取的"北纬34.17'24.41",东经57.33'20.18""周边的地形。从这里,我们对所看到的拥有空间意义的部分进行抽取,并对其进行空间图式层面的转化(图2-15-2)。在此基础上,我们进一步对所获得的空间图式进行空间化处理,便可以获得拥有建筑空间意义的形态(图2-15-3)。图2-15-4至图2-15-6是所获得的空间形态在不同视角下的空间与形态的呈现。

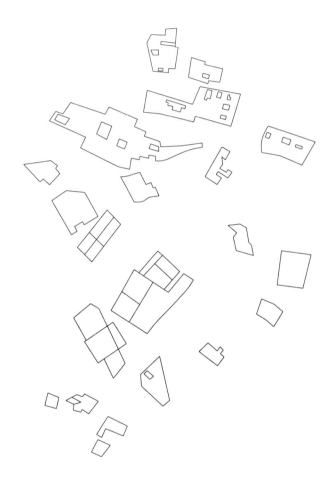

图2-15-2 从"北纬34.17'24.41",东经57.33'20.18""周边地形中所选取出的空间图式
Figure 2-15-2 Spatial diagrams chosen from landform around 34.17'24.41" N, 57.33'20.18" E

Figure 2-15-1 on the left page shows landform around 34.17'24.41" N, 57.33'20.18" E extracted from electronic maps. We extract parts with spatial meaning and conduct schematic transformation (figure 2-15-2). Based on this, we further conduct spatial processing on acquired spatial diagrams to obtain patterns with architectural spatial meaning (figure 2-15-3). Figure 2-15-4 to figure 2-15-6 represent space and pattern of acquired spatial patterns under different perspectives.

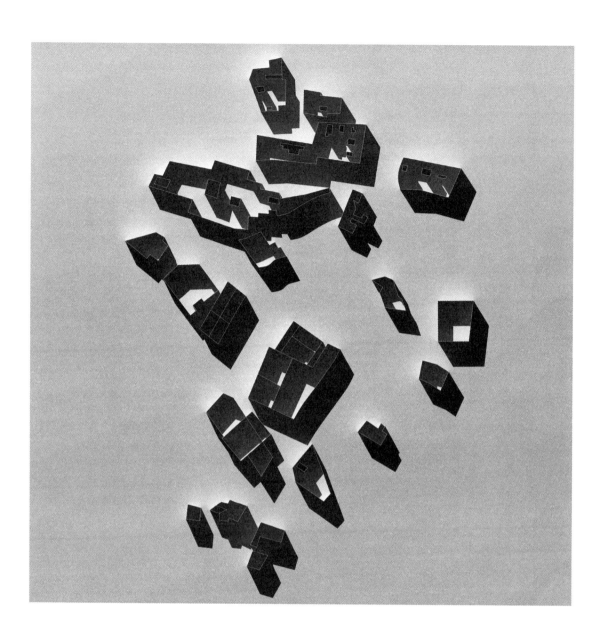

图2-15-3 根据空间图式所获得的空间与建筑形态
Figure 2-15-3 Spatial and architectural patterns acquired from spatial diagram

图2-15-4 空间与建筑形态在不同视角下的呈现
Figure 2-15-4 Space and architectural patterns under different perspectives

图2-15-5 空间与建筑形态在不同视角下的呈现
Figure 2-15-5 Space and architectural patterns under different perspectives

图2-15-6 空间与建筑形态在不同视角下的呈现
Figure 2-15-6 Space and architectural patterns under different perspectives

实例2-16 从"北纬39.59'50.83",东经120.43'55.90""周边地形所选择出的建筑
Sample 2-16 Architecture chosen from landform around 39.59'50.83" N, 120.43'55.90" E

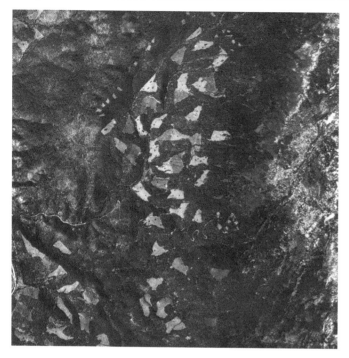

图2-16-1 "北纬39.59'50.83",东经120.43'55.90""周边地形卫星图
Figure 2-16-1 Satellite map of landform around 39.59'50.83" N, 120.43'55.90" E

图2-16-1所呈现的是从电子地图中截取的"北纬39.59'50.83″,东经120.43'55.90″"周边的地形。从这里,我们对所看到的拥有空间意义的部分进行抽取,并对其进行空间图式层面的转化(图2-16-2)。在此基础上,我们进一步对所获得的空间图式进行空间化处理,便可以获得拥有建筑空间意义的形态(图2-16-3)。图2-16-4和图2-16-5是所获得的空间形态在不同视角下的空间与形态的呈现。

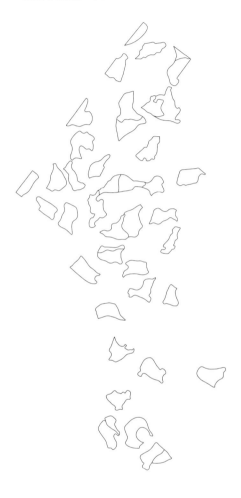

图2-16-2 从"北纬39.59'50.83″,东经120.43'55.90″"周边地形中所选取出的空间图式
Figure 2-16-2 Spatial diagrams chosen from landform around 39.59'50.83″ N, 120.43'55.90″ E

Figure 2-16-1 on the left page shows landform around 39.59'50.83" N, 120.43'55.90" E extracted from electronic maps. We extract parts with spatial meaning and conduct schematic transformation (figure 2-16-2). Based on this, we further conduct spatial processing on acquired spatial diagrams to obtain patterns with architectural spatial meaning (figure 2-16-3). Figure 2-16-4 and figure 2-16-5 represent space and pattern of acquired spatial patterns under different perspectives.

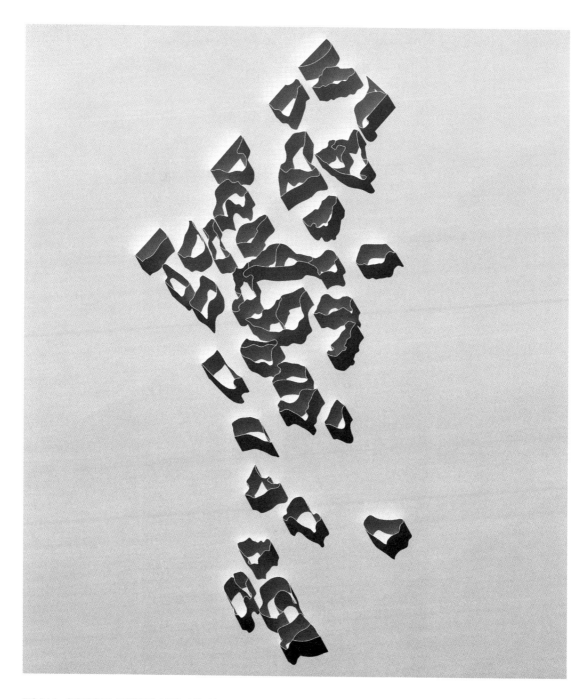

图2-16-3 根据空间图式所获得的空间与建筑形态

Figure 2-16-3 Spatial and architectural patterns acquired from spatial diagram

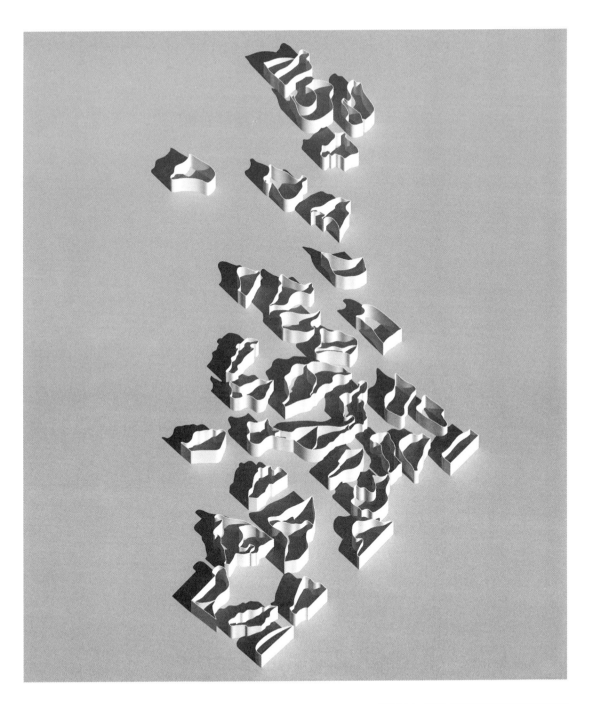

图2-16-4 空间与建筑形态在不同视角下的呈现
Figure 2-16-4 Space and architectural patterns under different perspectives

图2-16-5 空间与建筑形态在不同视角下的呈现
Figure 2-16-5 Space and architectural patterns under different perspectives

实例2-17 从"北纬37.30'33.17",东经122.00'32.90""周边地形所选择出的建筑
Sample 2-17 Architecture chosen from landform around 37.30'33.17" N, 122.00'32.90" E

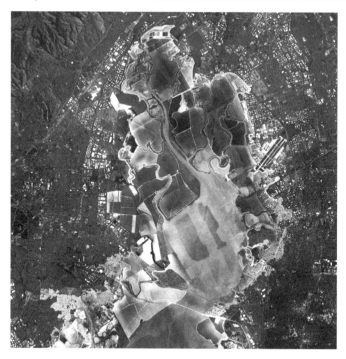

图2-17-1 "北纬37.30'33.17",东经122.00'32.90""周边地形卫星图
Figure 2-17-1 Satellite map of landform around 37.30'33.17" N, 122.00'32.90" E

图2-17-1所示呈现的是从电子地图中截取的"北纬37.30'33.17",东经122.00'32.90""周边的地形。从这里,我们对所看到的拥有空间意义的部分进行抽取,并对其进行空间图式层面的转化(图2-17-2)。在此基础上,我们进一步对所获得的空间图式进行空间化处理,便可以获得拥有建筑空间意义的形态(图2-17-3)。图2-17-4和图2-17-5是所获得的空间形态在不同视角下的空间与形态的呈现。

图2-17-2 从"北纬37.30'33.17",东经122.00'32.90""周边地形中所选取出的空间图式
Figure 2-17-2 Spatial diagrams chosen from landform around 37.30'33.17" N, 122.00'32.90" E

Figure 2-17-1 on the left page shows landform around 37.30'33.17" N, 122.00'32.90" E extracted from electronic maps. We extract parts with spatial meaning and conduct schematic transformation (figure 2-17-2). Based on this, we further conduct spatial processing on acquired spatial diagrams to obtain patterns with architectural spatial meaning (figure 2-17-3). Figure 2-17-4 and figure 2-17-5 represent space and pattern of acquired spatial patterns under different perspectives.

图2-17-3 根据空间图式所获得的空间与建筑形态
Figure 2-17-3 Spatial and architectural patterns acquired from spatial diagram

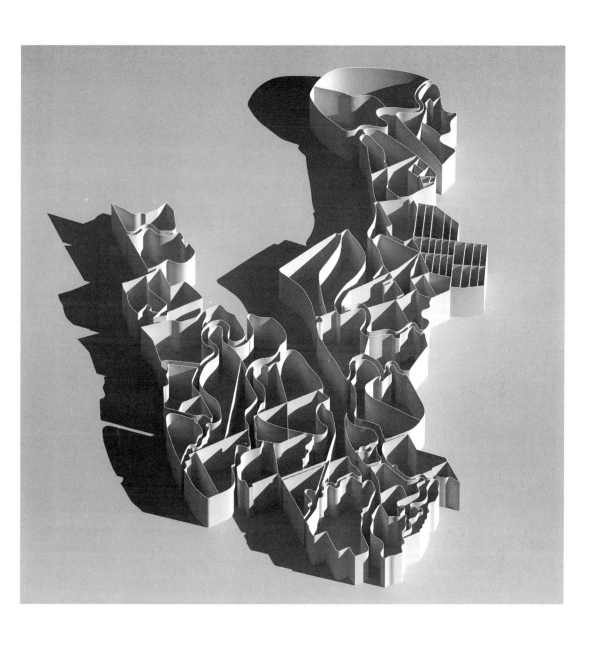

图2-17-4 空间与建筑形态在不同视角下的呈现
Figure 2-17-4 Space and architectural patterns under different perspectives

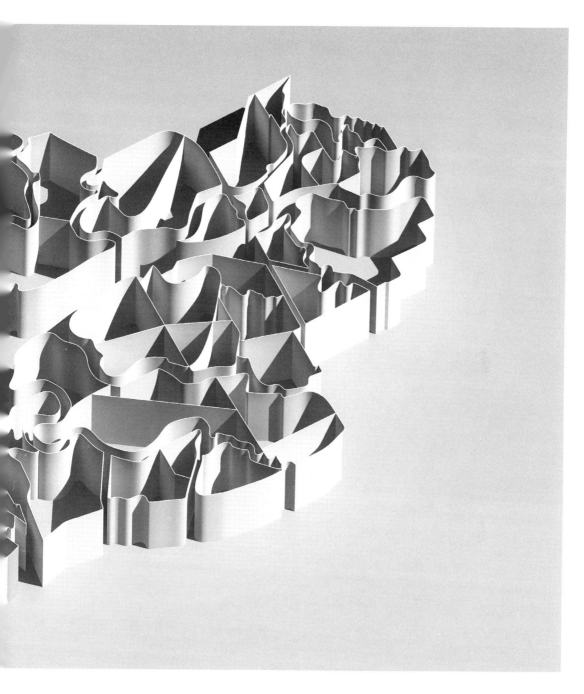

图2-17-5 空间与建筑形态在不同视角下的呈现
Figure 2-17-5 Space and architectural patterns under different perspectives

实例2-18 从"北纬35.38'12.54",东经120.46'43.01""周边地形所选择出的建筑
Sample 2-18 Architecture chosen from landform around 35.38'12.54" N, 120.46'43.01" E

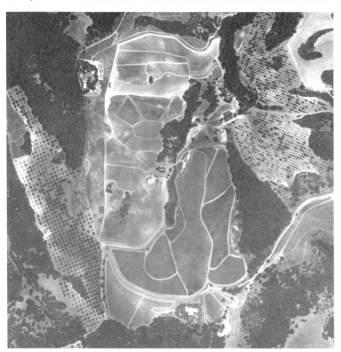

图2-18-1 "北纬35.38'12.54",东经120.46'43.01""周边地形卫星图
Figure 2-18-1 Satellite map of landform around 35.38'12.54" N, 120.46'43.01" E

图2-18-1所呈现的是从电子地图中截取的"北纬35.38'12.54",东经120.46'43.01""周边的地形。从这里,我们对所看到的拥有空间意义的部分进行抽取,并对其进行空间图式层面的转化(图2-18-2)。在此基础上,我们进一步对所获得的空间图式进行空间化处理,便可以获得拥有建筑空间意义的形态(图2-18-3)。图2-18-4和图2-18-5是所获得的空间形态在不同视角下的空间与形态的呈现。

图2-18-2 从"北纬35.38'12.54",东经120.46'43.01""周边地形中所选取出的空间图式
Figure 2-18-2 Spatial diagrams chosen from landform around 35.38'12.54" N, 120.46'43.01" E

Figure 2-18-1 on the left page shows landform around 35.38'12.54" N, 120.46'43.01" E extracted from electronic maps. We extract parts with spatial meaning and conduct schematic transformation (figure 2-18-2). Based on this, we further conduct spatial processing on acquired spatial diagrams to obtain patterns with architectural spatial meaning (figure 2-18-3). Figure 2-18-4 and figure 2-18-5 represent space and pattern of acquired spatial patterns under different perspectives.

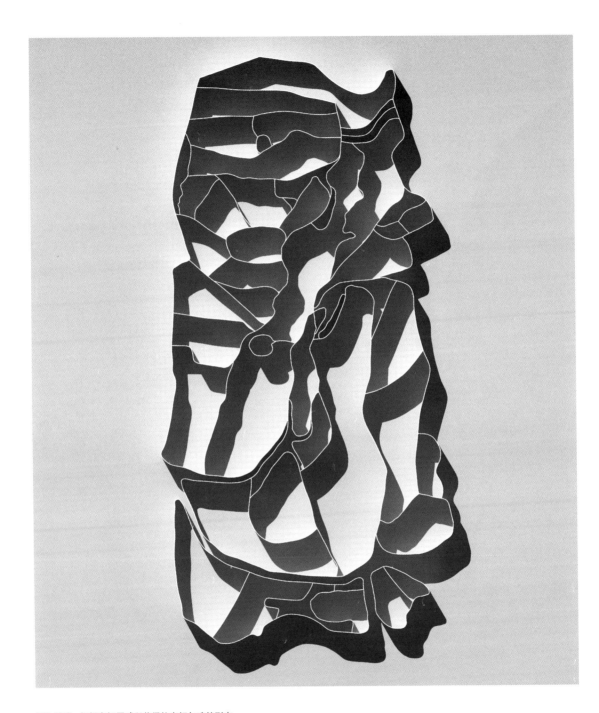

图2-18-3 根据空间图式所获得的空间与建筑形态
Figure 2-18-3 Spatial and architectural patterns acquired from spatial diagram

图2-18-4 空间与建筑形态在不同视角下的呈现
Figure 2-18-4 Space and architectural patterns under different perspectives

图2-18-5 空间与建筑形态在不同视角下的呈现
Figure 2-18-5 Space and architectural patterns under different perspectives

实例2-19 从"北纬40.45'16.15",东经144.52'45.92""周边地形所选择出的建筑
Sample 2-19 Architecture chosen from landform around 40.45'16.15" N, 144.52'45.92" E

图2-19-1 "北纬40.45'16.15",东经144.52'45.92""周边地形卫星图
Figure 2-19-1 Satellite map of landform around 40.45'16.15" N, 144.52'45.92" E

图2-19-1所呈现的是从电子地图中截取的"北纬40.45'16.15",东经144.52'45.92""周边的地形。从这里,我们对所看到的拥有空间意义的部分进行抽取,并对其进行空间图式层面的转化(图2-19-2)。在此基础上,我们进一步对所获得的空间图式进行空间化处理,便可以获得拥有建筑空间意义的形态(图2-19-3)。图2-19-4和图2-19-5是所获得的空间形态在不同视角下的空间与形态的呈现。

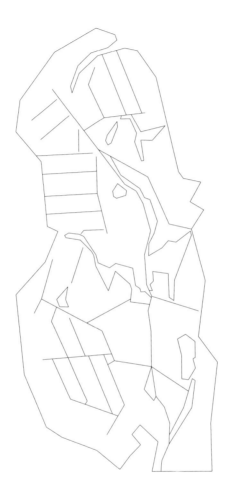

图2-19-2 从"北纬40.45'16.15",东经144.52'45.92""周边地形中所选取出的空间图式
Figure 2-19-2 Spatial diagrams chosen from landform around 40.45'16.15" N, 144.52'45.92" E

Figure 2-19-1 on the left page shows landform around 40.45'16.15" N, 144.52'45.92" E extracted from electronic maps. We extract parts with spatial meaning and conduct schematic transformation (figure 2-19-2). Based on this, we further conduct spatial processing on acquired spatial diagrams to obtain patterns with architectural spatial meaning (figure 2-19-3). Figure 2-19-4 and figure 2-19-5 represent space and pattern of acquired spatial patterns under different perspectives.

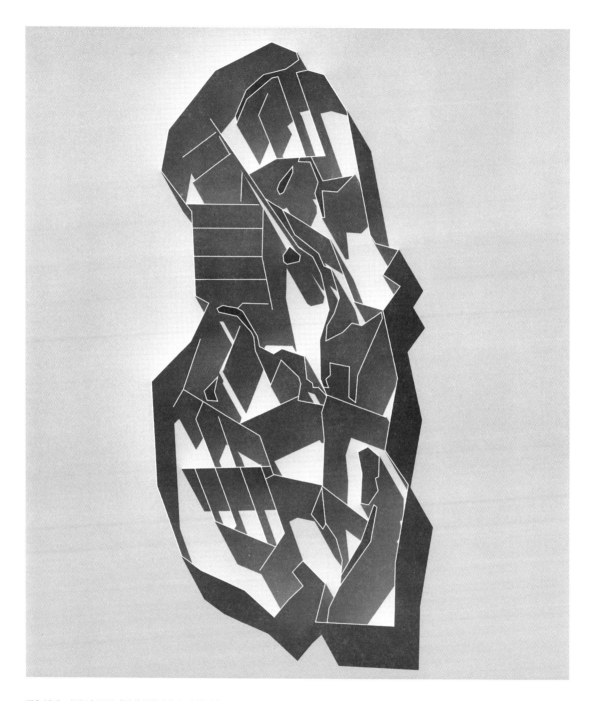

图2-19-3 根据空间图式所获得的空间与建筑形态

Figure 2-19-3 Spatial and architectural patterns acquired from spatial diagram

图2-19-4 空间与建筑形态在不同视角下的呈现

Figure 2-19-4 Space and architectural patterns under different perspectives

图2-19-5 空间与建筑形态在不同视角下的呈现
Figure 2-19-5 Space and architectural patterns under different perspectives

上一章我们从"肌理"的视角,对自然的形态进行了一系列试做,接下来,我们将对由地形隆起所产生的空间与建筑间的相互所指关系进行一系列陈述。地形是自然中的一部分,地形地貌被固化下来的本身呈现着一种由于力学平衡所造就的富有美感的形态与造型。我们从大自然中任意地选择一块地形,在电子地图上进行直接截取,从而可以直接生成具有当代性的造型。本章节中,我们将列举5个案例,以示例从自然形态中,通过直接截取而获得的"建筑"。

In the previous chapter, we conducted a series of experiments on natural patterns from the perspective of "texture". Next, we will offer a series of narration on mutual relationship between architecture and space generated by landform upheavals. Landforms in solidification, as part of nature, reveal patterns and shapes full of beauty caused by mechanical equilibrium. We choose a landform randomly from nature and directly extract it from electronic maps, leading to shapes of contemporary sense straightly. In this chapter, we will offer 5 samples to illustrate "architecture" directly extracted from natural patterns.

从自然中直接截取的建筑
Architecture directly extracted from nature

实例3-1 直接截取"北纬36.75'84.63"，东经110.49'35""周边隆起的地形所获得的空间形态

Sample 3-1 Spatial patterns acquired by directly extracting upheaval area around 36.75'84.63"N, 110.49'35" E

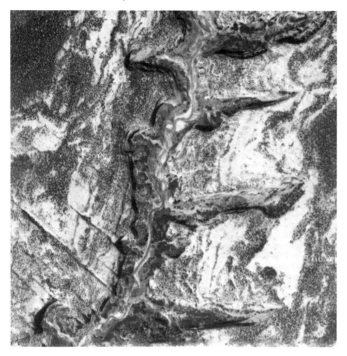

图3-1-1 "北纬36.75'84.63"，东经110.49'35""周边地形卫星图
Figure 3-1-1 Satellite map of landform around 36.75'84.63"N, 110.49'35" E

图3-1-1是从电子地图中直接截取的"北纬36.75'84.63",东经110.49'35""周边地形,我们通过操作软件进行处理,产生出图3-1-2所呈现的形态。进一步,我们对其隆起的形态直接进行空间化处理,便产生了具有建筑指向的空间形态(图3-1-3和图3-1-4),同时也获得了其在不同角度下的空间场景(图3-1-5和图3-1-6)。

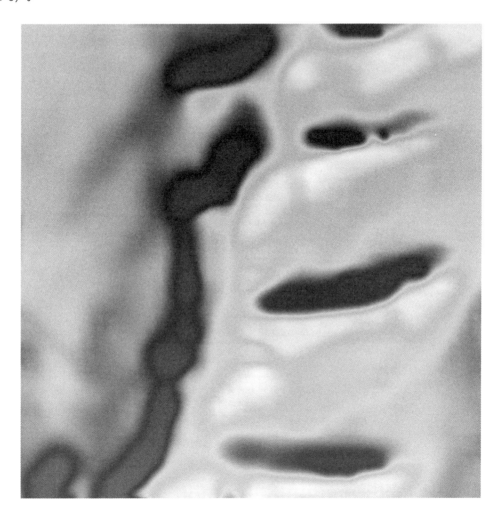

图3-1-2 从"北纬36.75'84.63",东经110.49'35""周边地形中直接截取的空间俯视图
Figure 3-1-2 Aerial view directly extracted from landform around 36.75'84.63"N, 110.49'35" E

Figure 3-1-1 on the left page shows landform around 36.75'84.63"N, 110.49'35" E directly extracted from electronic maps. We use software processing to produce the patterns shown by figure 3-1-2. Furthermore, we directly conduct spatial processing on upheaval patterns so as to produce spatial patterns with architectural indication (figure 3-1-3 and figure 3-1-4) as well as acquire spatial scenes under different perspectives (figure 3-1-5 and figure 3-1-6).

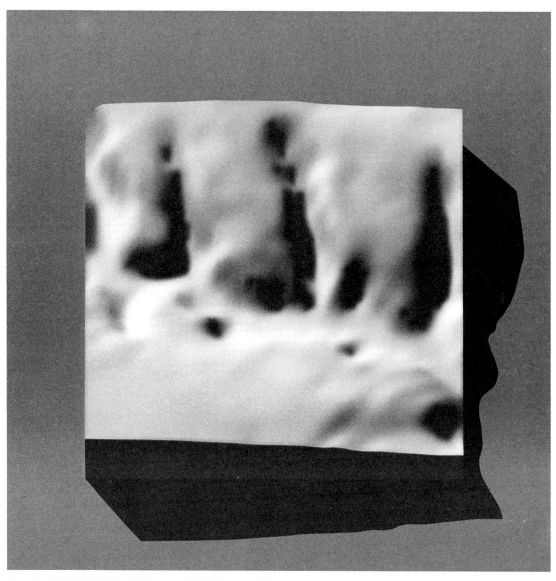

图3-1-3 对直接截取后所获得的隆起的形态进行空间化处理后所获得的"建筑"形态(西侧角度)
Figure 3-1-3 "Architectural patterns" acquired after spatial processing on upheaval patterns extracted directly (west angle)

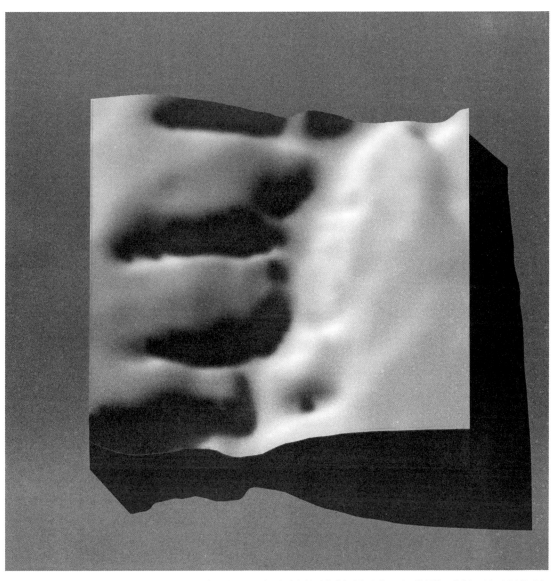

图3-1-4 对直接截取后所获得的隆起的形态进行空间化处理后所获得的"建筑"形态（北侧角度）
Figure 3-1-4 "Architectural patterns" acquired after spatial processing on upheaval patterns extracted directly (north angle)

图3-1-5 对直接截取后所获得的隆起的形态进行空间化处理后所获得的"建筑"形态（东立面）
Figure 3-1-5 "Architectural patterns" acquired after spatial processing on upheaval patterns extracted directly (east facade)

图3-1-6 对直接截取后所获得的隆起的形态进行空间化处理后所获得的"建筑"形态(北立面)
Figure 3-1-6 "Architectural patterns" acquired after spatial processing on upheaval patterns extracted directly (north facade)

实例3-2 直接截取"北纬37.46'65.09″,东经110.99'91.12′″"周边隆起的地形所获得的正向空间形态

Sample 3-2 Spatial patterns of forward direction acquired from upheaval landform around 37.46'65.09" N, 110.99'91.12" E

图3-2-1 "北纬37.46'65.09″,东经110.99'91.12′″"周边地形卫星图
Figure 3-2-1 Satellite map of landform around 37.46'65.09" N, 110.99'91.12" E

图3-2-1是从电子地图中直接截取的"北纬37.46'65.09"，东经110.99'91.12'""周边地形，我们通过操作软件进行处理，产生出图3-1-2所呈现的正向形态。进一步地，我们对其隆起的形态直接进行空间化处理，便产生了具有建筑指向的正向空间形态（图3-2-3和图3-2-4），同时也获得了其在不同角度下的空间场景（图3-2-5和图3-2-6）。

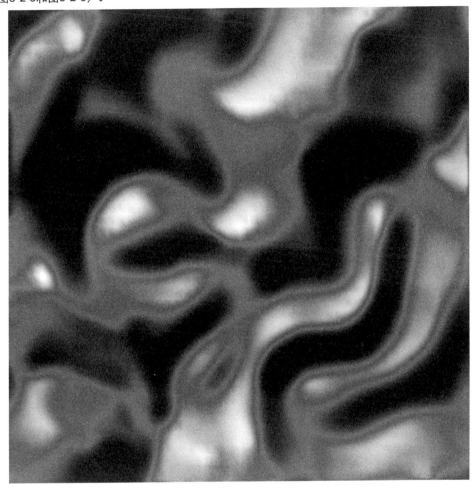

图3-2-2 从"北纬37.46'65.09"，东经110.99'91.12'""周边地形中直接截取的正向空间俯视图
Figure 3-2-2 Space aerial view of forward direction directly extracted from landform around 37.46'65.09" N, 110.99'91.12" E

Figure 3-2-1 on the left page shows landform around 37.46'65.09" N, 110.99'91.12" E directly extracted from electronic maps. We use software processing to produce the forward patterns shown by figure 3-1-2. Furthermore, we directly conduct spatial processing on upheaval patterns so as to produce positive spatial patterns with architectural indication (figure 3-2-3 and figure 3-2-4), as well as aquire spatial scenes under different perspectives (figure 3-2-5 and figure 3-2-6).

图3-2-3 对直接截取后所获得的隆起的形态进行空间化处理后所获得的"建筑"的形态（北侧角度）
Figure 3-2-3 "Architectural patterns" acquired after spatial processing on upheaval patterns extracted directly (north angle)

图3-2-4 对直接截取后所获得的隆起的形态进行空间化处理后所获得的"建筑"的形态(西侧角度)
Figure 3-2-4 "Architectural patterns" acquired after spatial processing on upheaval patterns extracted directly (west angle)

图3-2-5 对直接截取后所获得的隆起的形态进行空间化处理后所获得的"建筑"的形态(东立面)
Figure 3-2-5 "Architectural patterns" acquired after spatial processing on upheaval patterns extracted directly (east facade)

图3-2-6 对直接截取后所获得的隆起的形态进行空间化处理后所获得的"建筑"的形态(南立面)
Figure 3-2-6 "Architectural patterns" acquired after spatial processing on upheaval patterns extracted directly (south facade)

实例3-3 直接截取"北纬37.46'65.09'',东经110.99'91.12'""周边隆起的地形所获得的负向空间形态

Sample 3-3 Spatial patterns of negative direction acquired from upheaval landform around 37.46'65.09" N, 110.99'91.12" E

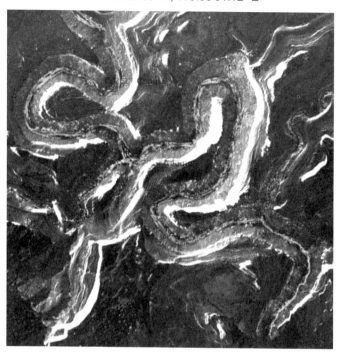

图3-3-1 "北纬37.46'65.09'',东经110.99'91.12'""周边地形卫星图
Figure 3-3-1 Satellite map of landform around 37.46'65.09" N, 110.99'91.12" E

图3-3-1是从电子地图中直接截取的"北纬37.46'65.09'',东经110.99'91.12'"周边地形,我们通过操作软件进行处理,产生出图3-3-2所呈现的负向形态。进一步地,我们对其隆起的形态直接进行空间化处理,便产生了具有建筑指向的负向空间形态(图3-3-3和图3-3-4),同时也获得了其在不同角度下的空间场景(图3-3-5和图3-3-6)。

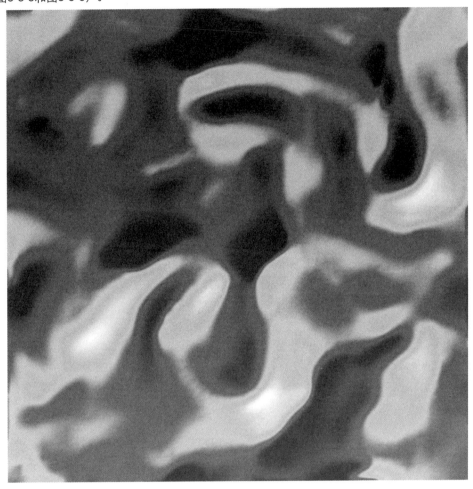

图3-3-2 从"北纬37.46'65.09'',东经110.99'91.12'"周边地形中直接截取的负向空间俯视图
Figure 3-3-2 Negative space aerial view directly extracted from landform around 37.46'65.09" N, 110.99'91.12" E

Figure 3-3-1 on the left page shows landform around 37.46'65.09" N, 110.99'91.12" E directly extracted from electronic maps. We use software processing to produce the negative patterns shown by figure 3-3-2. Furthermore, we directly conduct spatial processing on upheaval patterns so as to produce negative spatial patterns with architectural indication (figure 3-3-3 and figure 3-3-4), as well as acquire spatial scenes under different perspectives (figure 3-3-5 and figure 3-3-6).

图3-3-3 对直接截取后所获得的隆起的形态进行空间化处理后所获得的"建筑"的形态(北侧角度)
Figure 3-3-3 "Architectural patterns" acquired after spatial processing on upheaval patterns extracted directly (north angle)

图3-3-4 对直接截取后所获得的隆起的形态进行空间化处理后所获得的"建筑"的形态(西侧角度)
Figure 3-3-4 "Architectural patterns" acquired after spatial processing on upheaval patterns extracted directly (west angle)

图3-3-5 对直接截取后所获得的隆起的形态进行空间化处理后所获得的"建筑"的形态(东立面)
Figure 3-3-5 "Architectural patterns" acquired after spatial processing on upheaval patterns extracted directly (east facade)

图3-3-6 对直接截取后所获得的隆起的形态进行空间化处理后所获得的"建筑"的形态(南立面)
Figure 3-3-6 "Architectural patterns" acquired after spatial processing on upheaval patterns extracted directly (south facade)

实例3-4 直接截取"北纬24.38'16.94",东经121.11'84.74'"周边隆起的地形所获得的空间形态

Sample 3-4 Spatial patterns acquired by directly extracting upheaval area around 24.38'16.94"N, 121.11'84.74" E

图3-4-1 "北纬24.38'16.94",东经121.11'84.74'"周边地形卫星图
Figure 3-4-1 Satellite map of landform around 24.38'16.94"N, 121.11'84.74" E

图3-4-1是从电子地图中直接截取的"北纬24.38'16.94'',东经121.11'84.74'''"周边地形,我们通过操作软件进行处理,产生出图3-1-2所呈现的形态。进一步地,我们对其隆起的形态直接进行空间化处理,便产生了具有建筑指向的空间形态(图3-4-3和图3-4-4)。

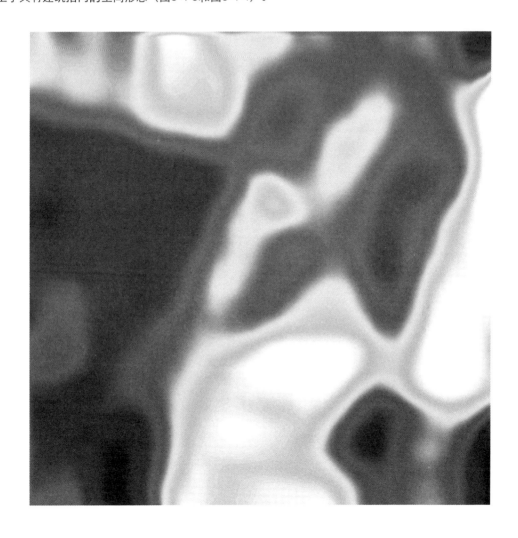

图3-4-2 从"北纬24.38'16.94'',东经121.11'84.74'''"周边地形中直接截取的空间俯视图
Figure 3-4-2 Aerial view directly extracted from landform around 24.38'16.94"N, 121.11'84.74" E

Figure 3-4-1 on the left page shows landform around 24.38'16.94"N, 121.11'84.74" E directly extracted from electronic maps. We use software processing to produce the patterns shown by figure 3-4-2. Furthermore, we directly conduct spatial processing on upheaval patterns so as to produce spatial patterns with architectural indication (figure 3-4-3 and figure 3-4-4).

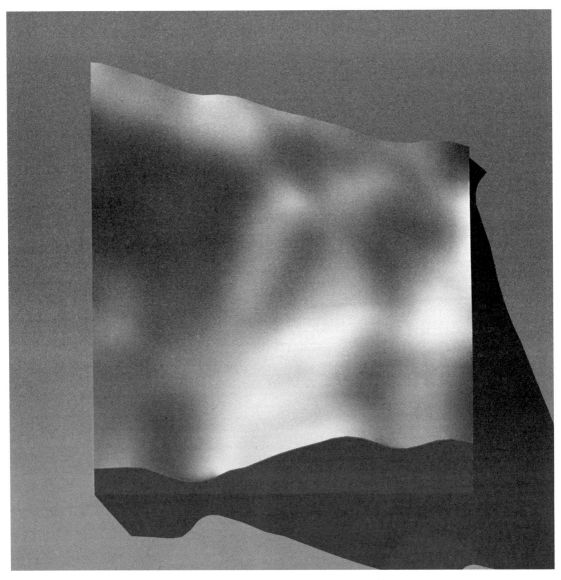

图3-4-3 对直接截取后所获得的隆起的形态进行空间化处理后所获得的"建筑"的形态(南侧角度)
Figure 3-4-3 "Architectural patterns" acquired after spatial processing on upheaval patterns extracted directly (south angle)

图3-4-4 对直接截取后所获得的隆起的形态进行空间化处理后所获得的"建筑"的形态(东侧角度)
Figure 3-4-4 "Architectural patterns" acquired after spatial processing on upheaval patterns extracted directly (east angle)

实例3-5 直接截取"北纬7.32'52.19″,东经134.49'15.70'″"周边隆起的地形所获得的空间形态

Sample 3-5 Spatial patterns acquired by directly extracting upheaval area around 7.32'52.19"N, 134.49'15.70" E

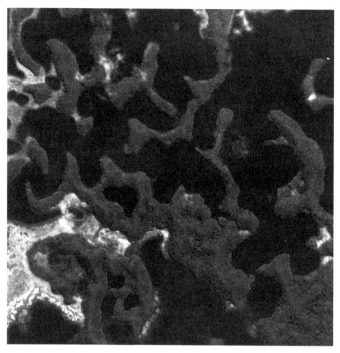

图3-5-1 "北纬7.32'52.19″,东经134.49'15.70'″" 周边地形卫星图
Figure 3-5-1 Satellite map of landform around 7.32'52.19"N, 134.49'15.70" E

图3-5-1是从电子地图中直接截取的"北纬7.32'52.19"，东经134.49'15.70'"周边地形，我们通过操作软件进行处理，产生出图3-5-2所呈现的形态。进一步地，我们对其隆起的形态直接进行空间化处理，便产生了具有建筑指向的空间形态（图3-5-3至图3-5-6），同时也获得了其在不同角度下的空间场景（图3-5-7）。

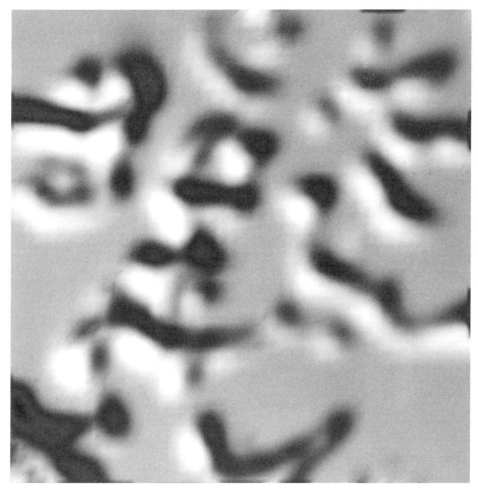

图3-5-2　从"北纬7.32'52.19"，东经134.49'15.70'"周边地形中直接截取的空间俯视图
Figure 3-5-2 Aerial view directly extracted from landform around 7.32'52.19"N, 134.49'15.70" E

Figure 3-5-1 on the left page shows landform around 7.32'52.19"N, 134.49'15.70" E directly extracted from electronic maps. We use software processing to produce the patterns shown by figure 3-5-2. Furthermore, we directly conduct spatial processing on upheaval patterns so as to produce spatial patterns with architectural indication (figure 3-5-3 to figure 3-5-6) as well as acquire spatial scenes under different perspectives (figure 3-5-7).

图3-5-3 对直接截取后所获得的隆起的形态进行空间化处理后所获得的"建筑"的形态(西侧角度)
Figure 3-5-3 "Architectural patterns" acquired after spatial processing on upheaval patterns extracted directly (west angle)

作者简介
About the author

王昀简介

王昀 博士

1985 年毕业于北京建筑工程学院建筑系
　　　获学士学位
1995 年毕业于日本东京大学
　　　获工学硕士学位
1999 年毕业于日本东京大学
　　　获工学博士学位
2001 年执教于北京大学
2002 年成立方体空间工作室
2013 年创立北京建筑大学建筑设计艺术研究中心
　　　担任主任
2015 年于清华大学建筑学院担任设计导师

建筑设计竞赛获奖经历：
1993 年日本《新建筑》第 20 回日新工业建筑设计
　　　竞赛获二等奖
1994 年日本《新建筑》第 4 回 S×L 建筑设计竞赛
　　　获一等奖

主要建筑作品：
善美办公楼门厅增建，60 ㎡极小城市，石景山财政局培训中心，庐师山庄，百子湾中学，百子湾幼儿园，杭州西溪湿地艺术村 H 地块会所等。

参加展览：
2004 年 6 月 "'状态'中国青年建筑师 8 人展"
2004 年首届中国国际建筑艺术双年展
2006 年第二届中国国际建筑艺术双年展
2009 年比利时布鲁塞尔 "'心造'——中国当代建筑
　　　前沿展"
2010 年威尼斯建筑艺术双年展，德国卡尔斯鲁厄
　　　Chinese Regional Architectural Creation
　　　建筑展
2011 年捷克布拉格中国当代建筑展，意大利罗马
　　　"向东方——中国建筑景观"展，中国深圳·香港
　　　城市建筑双城双年展
2012 年第十三届威尼斯国际建筑艺术双年展中国馆等

Wang Yun Profile

Dr. Wang Yun

Graduated with a Bachelor's degree from the Department of Architecture at the Beijing Institute of Architectural Engineering in 1985.
Received his Master's degree in Engineering Science from Tokyo University in 1995.
Received a Ph.D. from Tokyo University in 1999.
Taught at Peking University since 2001.
Founded the Atelier Fronti (www.fronti.cn) in 2002.
Established Graduate School of Architecture Design and Art of Beijing University of Civil Engineering
and Architecture in 2013, served as dean.
Served as a design instructor at School of Architecture, Tsinghua University in 2015.

Prize:
Received the second place prize in the "New Architecture" category at Japan's 20th annual International Architectural Design Competition in 1993.
Awarded the first prize in the "New Architecture" category at Japan's 4th SxL International Architectural Design Competition in 1994.

Prominent works:
ShanMei Office Building Foyer, a Small City of 60 Square Meters, the Shijingshan Bureau of Finance Training Center, Lushi Mountain Villa, Baiziwan Middle School, Baiziwan Kindergarten, and Block H of the Hangzhou Xixi Wetland Art Village.

Exhibitions:
The 2004 Chinese National Young Architects 8 Man Exhibition, the First China International Architecture Biennale, the Second China International Architecture Biennale in 2006, the "Heart-Made: Cutting-Edge of Chinese Contemporary Architecture" exhibit in Brussels in 2009, the 2010 Architectural Venice Biennale, the Karlsruhe Chinese Regional Architectural Creation exhibition in Germany, the Chinese Contemporary Architecture Exhibition in Prague in 2011, the "Towards the East: Chinese Landscape Architecture" exhibition in Rome, the Hong Kong-Shenzhen Twin Cities Urban Planning Biennale, Pavilion of China The 13th international Architecture Exhibition la Biennale di Venezia in 2012.

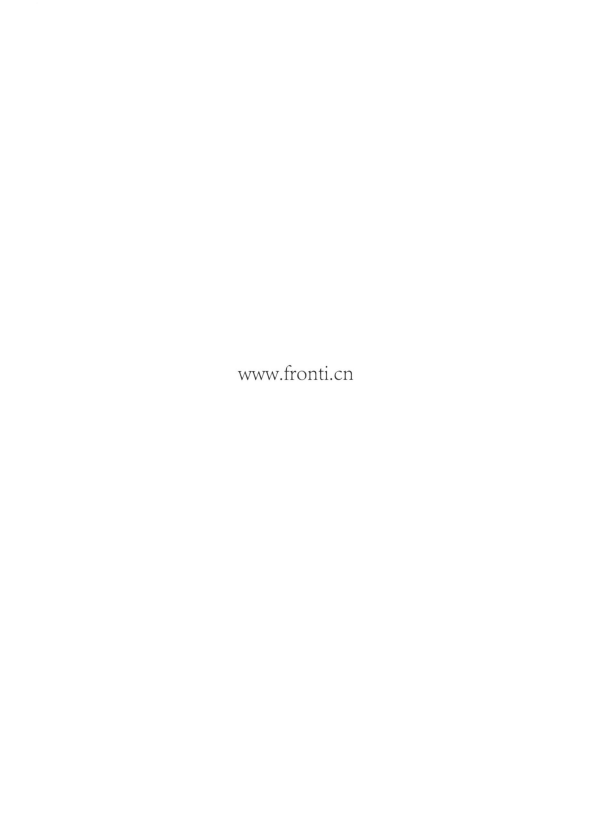